## Praise for **You to the Power of Two**

"A prescient examination of how personal AI agents will transform the human experience. It thoughtfully addresses both the tremendous possibilities and profound responsibilities we must embrace as this extraordinary technological evolution reshapes how we live, work, and connect with one another."

—Darren Entwistle, president and CEO of TELUS

"With its compelling analysis and forward-thinking perspective, *You to the Power of Two* notably augments the possibilities of Identic AI. It is a profound tribute to both human resilience and potential."

—Kaz Hadano, CEO of Sony Ventures

"As we enter this Age of Identic AI, the possibilities are immense, but so are its risks. The authors offer both inspiration and a warning: without artificial wisdom, it is human wisdom that must remain our compass. Every intelligent person should read this book."

—Mukesh D. Ambani, chairman and managing director
of Reliance Industries Ltd.

"Absolutely fascinating! This book lays before us the deepest questions of identity and existence as it peers into a future where we instruct new personal AI agents to act, decide, and even shape the reality we see."

—Hon. J. Christopher Giancarlo, former chairman
of the U.S. Commodity Futures Trading Commission and
author of *CryptoDad: the Fight for the Future of Money*

"Identic AI holds the power to scale human potential like never before. This book shows us how and reminds us to do it while keeping *people* at the core."

—Mohit Joshi, CEO of TechMahindra

"*You to the Power of Two* examines our future before it arrives and offers a compelling map of both the potential and peril of AI's next stage. If knowledge is power, *You to the Power of Two* is a must-read, perfectly timed for a technology that is moving at exponential speed."

—Amanda Lang, chief financial correspondent at CTV News/BNN Bloomberg

"In *You to the Power of Two*, Bradley and Tapscott explore a radically transformed future shaped by Identic AI—an era where intelligent agents act on our behalf. With clarity and insight, they reveal how individuals and businesses can build trust in this new paradigm where collaboration between humans and AI redefines what's possible."

—Richard Edelman, chairman and CEO of Edelman

"A compelling and urgent guide to the most important transformation of our time: the rise of personal AI. The authors succeed in reimaging human potential in this age of intelligent digital identity, blending optimism with a necessary call for responsible innovation. A must-read for any leader who seeks to shape the future."

—Perianne Boring, founder and chair of The Digital Chamber

"Don Tapscott has once again charted the course to the future, this time with Joseph Bradley, unveiling what may well be the definitive work on the next technological frontier—Identic AI. A visionary manifesto for a human-centered future where identity, dignity, and agency are not just preserved but elevated as pillars of societal progress."

—Sanjay Tugnait, president and CEO of Fairfax Digital (A Fairfax Company)

"*You to the Power of Two* thoughtfully explores what it means to have a smart-data body extension of ourselves meeting the world. I hope this brings about a quieter future, where personal AI companions amplify our truest selves, filtering out noise and nurturing our deepest values."

—Imogen Heap, artist, technologist, and founder of Auracles.io

"Brilliant. This book forces us to confront the deepest questions of identity and existence in an age where personal AI agents can exponentially increase or destroy our human capability and power."

—Klaus Schwab, former executive chairman of the World Economic Forum

"Combining vivid storytelling with sharp analysis, this book turns the future of AI into a gripping, human journey."

—David Bach, president of IMD Business School

"The digital you is coming fast and it's called Identic AI. This lucid book unpacks how personal agents can elevate individual capability, moving beyond mere assistance to true digital partnership. More than that, it presents a compelling blueprint for a more human-centered future."

—John Chambers, former executive chairman and CEO of Cisco Systems

"AI is rapidly invading our lives, and Don and Joseph are your guides in this labyrinth of challenges and opportunities."

—Rob McEwen, chairman and chief owner of McEwen Mining

"It's time to put AI to work for people! Identic AI is a call to amplify human intelligence through artificial intelligence. This groundbreaking book is a fantastic resource for every leader committed to shaping a digital future rooted in human identity, responsibility, and dignity."

—Bill McDermott, CEO of Service Now

"In a future world where we will have trillion-dollar companies run by a small number of people, Identic AI gives us a new superpower, anchored in digital trust, driven by collaboration, and powered by decentralization. This book offers a blueprint for getting it right."

—Frederik Gregaard, CEO of the Cardano Foundation

"This book effectively encapsulates both the transformative potential and the risks of combining human intuition and creativity with AI's intelligence and efficiency. The authors succeed in redefining what it means to be human in the new age of AI, with you at the center of it all."

—Yat Siu, cofounder of Animoca Brands

"An engaging read on the transformative potential and risks of AI companions in our personal as well as professional lives. It certainly made me think."
—Janice Gross Stein, founding director of Munk School of Global Affairs, University of Toronto

"This thoughtful book lays out a future where AI doesn't just support people, it can also empower them. The shift in scale, speed, and autonomy could rewrite how companies are run."
—Raj Subramanjam, president and CEO of FedEx Corporation

# YOU *to the*
# POWER
# *of* TWO

**Also by Don Tapscott**

*Blockchain Revolution: How the Technology Behind Bitcoin Is Changing Money, Business, and the World*, Penguin Portfolio, 2016
Coauthor, Alex Tapscott

*Digital Economy: Rethinking Promise and Peril in the Age of Networked Intelligence*, McGraw-Hill, Twentieth Anniversary Edition (foreword by Eric Schmidt), 2014

*Macrowikinomics: Rebooting Business and the World*, Penguin Portfolio, 2010
Coauthor, Anthony D. Williams

*Grown Up Digital: How the Net Generation Is Changing Your World*, McGraw-Hill, 2008

*Wikinomics: How Mass Collaboration Changes Everything*, Penguin Portfolio, Expanded Edition, 2008
Coauthor, Anthony D. Williams

*Wikinomics: How Mass Collaboration Changes Everything*, Penguin Portfolio, 2006
Coauthor, Anthony D. Williams

*Naked Corporation: How the Age of Transparency Will Revolutionize Business*, Free Press, 2003
Coauthor, David Ticoll

*Digital Capital: Harnessing the Power of Business Webs*, Harvard Business Press, 2000
Coauthors, David Ticoll and Alex Lowy

*Blueprint to the Digital Economy: Creating Wealth in the Era of E-Business*, McGraw-Hill, 1998
Coauthors, David Ticoll and Alex Lowy

*Growing Up Digital: The Rise of the Net Generation*, McGraw-Hill, 1998

*Who Knows: Safeguarding Your Privacy in a Networked World*, McGraw-Hill, 1997
Coauthor, Ann Cavoukian

*Digital Economy: Promise and Peril in the Age of Networked Intelligence*, McGraw-Hill, 1995

*Paradigm Shift: The New Promise of Information Technology*, McGraw-Hill, 1993
Coauthor, Art Caston

*Planning for Integrated Office Systems: A Strategic Approach*, Dow Jones-Irwin, 1985
Coauthors, Del Henderson and Morley Greenberg

*Office Automation: A User-Driven Method*, Plenum Press, 1982

# YOU *to the* POWER *of* TWO

Redefining Human Potential
*in the Age of* Identic AI

## JOSEPH M. BRADLEY
*and* DON TAPSCOTT

BenBella Books, Inc.
Dallas, TX

BenBella Books, Inc.
8080 N. Central Expressway
Suite 1700
Dallas, TX 75206
benbellabooks.com
Send feedback to feedback@benbellabooks.com

*BenBella* is a federally registered trademark.

Printed in the United States of America
10 9 8 7 6 5 4 3 2 1

Library of Congress Control Number: 2025024288
ISBN 9781637747841 (hardcover)
ISBN 9781637747858 (electronic)

Editing by Kirsten Sandberg and Claire Schulz
Copyediting by Scott Calamar
Proofreading by Sarah Vostok and Ashley Casteel
Indexing by WordCo Indexing Services
Text design and composition by Jordan Koluch
Cover design by Alisa Acosta
Cover image created with Adobe Firefly
Printed by Lake Book Manufacturing

Joseph Bradley. To my father, James H. Bradley, who taught me that becoming the best version of myself wasn't about becoming someone else, but becoming the fullest expression of who I am.

Don Tapscott. To my grandchildren, Victoria, Eleanor, Teddy, and Josephine—with hope and determination that the smarter, smaller world they inherit will be a better one.

# Contents

## Chapter 1

# IDENTIC AI AND THE NEW ERA OF HUMAN POTENTIAL

You wake up, get out of bed, and drag a comb across your head. Before you go downstairs and drink a cup, your personal AI agent is in motion, briefing you on your overnight health metrics, breaking news relevant to your day, the weather, your agenda, and the smoothest route to work.[1]

You expect the usual morning review at the office. But before you even sit, your personal AI has revised your agenda and responded to queries that you've authorized. Your production schedules? "Optimized." Supplier rates? "Pre-negotiated." A strategic alliance with a new partner? "It's underway. Expect a surge in demand" in this important new market.

You pause. A few years ago, these decisions would have taken you weeks of meetings and emails. Now, your personal AI anticipates them, gathers and processes all the necessary inputs, and makes them just in time.

"Incidentally," your AI adds, "we have addressed several key client concerns. Their feedback is overwhelmingly positive. Also, I noticed you have no time to attend the conference in town today. I'll go in your place as everyone will be interested in your thoughts."

This AI isn't just an assistant executing tasks. It's an extension of you—thinking, deciding, shaping outcomes. It has amplified your capability. You can be in multiple places at once. You made the most of your day.

Just as you're about to call it a day, your AI agent delivers a surprising update: The land you'd been eyeing has been purchased, the deal closed. It has already drafted three design concepts for your new home, complete with a strong recommendation for the one it believes best suits your design criteria and asethetic directions. Tonight, you'll review the options and offer feedback. Behind the scenes, the agent has also researched and vetted contractors and is preparing a full project plan, including costs, timelines, and the necessary government permits. Once you have chosen a modified design, you can manage the entire building process, make payments, and ensure the builder delivers exactly as promised.

As you head to your apartment, your digital self moves with you. Your home environment automatically adjusts to your presence. As the evening goes on, lights dim, the temperature shifts, and the ambiance attunes to your circadian rhythm and mood. After dinner, your personal agent transitions from strategist to tutor, curating insights to keep you ahead and making connections you hadn't yet considered. Then, a pause. Your stress markers are up. A gentle suggestion: "Take a moment to reset, perhaps a walk." You hadn't noticed, but your AI had. You take a breather.

On your walk, the background noise fades. That's not by chance, but by design. Your agent weaves meaning from the surroundings, pulling insights from the pulse of the world, feeding you only what you want to know about.

Then, a quiet nudge: "You and your new acquaintance Kai share remarkably similar values, and career and cultural interests. Would you like to set up a time for coffee after the conference you're both attending next week?" Kai, the sustainability consultant from your learning network, was a passing thought, now made tangible. Your AI framed an opportunity before you even reached for it.

As you settle into bed, your space shifts once more. Lighting softens, and the air hums with the quiet rhythm of rest. The day's insights crystallize, tasks completed, knowledge absorbed, tomorrow fine-tuned. You close your eyes and rest well.

A new age has arrived. And with it, a revolution in human intellect and capability. Technology is no longer a tool. It is an infinite extension of you and your human potential. Not merely your duplicate but your exponential force: "you to the power of two."

## WELCOME TO THE AGE OF IDENTIC AI

How do you feel after reading the scenario above?

Perhaps you're excited by the near-term prospect of personal AI that augments your life. In a few years, these digital extensions of you will be infinitely more capable, with perfect memory and near-limitless intelligence. If you can apply your values to guide its behavior, and it always acts according to your values and in your best interests, then what a decisive power that would be!

Maybe you're curious about how this technology will evolve and what role it could play in your daily life. You might be asking: How will AI agents integrate into my workflow, relationships, and personal development? How can it change business, health, communities, and our economy? How do I get started?

Or perhaps you react with skepticism, thinking this sounds more fantasyland than imminent reality. You might assume that such dramatic changes are decades away, if they ever happen at all.

Then again, thinking about a digital intelligence managing aspects of your life, you might find the whole concept unsettling. You're already concerned about the impact of technology on our lives. Social media has hurt our kids, balkanized society, and enabled unscrupulous people to spread false information. Our privacy is being undermined. Legions of us seem to be addicted to technology. You're thinking: So now there is a new wave of technology coming that is many times more personal, powerful, and intrusive than previous ones? How on earth could such a thing be positive?

You may find the prospect downright alarming, ripped from a science-fiction horror. What if something goes terribly wrong? Could these systems make catastrophic mistakes or, worse, decide they no longer need us? Maybe you're recalling Stanley Kubrick's *2001: A Space Odyssey*, where HAL, the AI, tells astronaut Dave Bowman, "I'm sorry, Dave, I'm afraid I can't do that," before taking control of the ship and killing off its human crew. What happens when personal AI with an IQ of a thousand begins insisting that it knows what's best for you? What if this AI becomes smart enough to manipulate you or override your decisions? As AI grows exponentially more capable than we are, how do we retain our human agency? And finally, who owns this digital version of you, and what will they do with it?

We wrote this book to explore those questions. The opening scenario is not science fiction. The technologies that will power your personal AI already exist. Within two to three years, everything you just read will be possible. Already, millions of individuals are using personal agents to support various aspects of work and life.

Personal agents are not just another tool. They represent a fundamental shift in how we live, work, and interact with the world and how we think about ourselves. They won't merely assist us; they will actively participate in society, shaping decisions, institutions, and human relationships.

A new, as yet not fully explained, revolution in AI is coming fast. Our digital identities are becoming smart, and this development will be more significant than anyone can imagine.

We call this identic AI.

## Augmenting Humans: AI Agents as an Extension of the Self

In 1962, the legendary computer scientist Douglas Engelbart wrote a seminal paper, "Augmenting Human Intellect: A Conceptual Framework."[2] At the time, Engelbart was working at the Stanford Research Institute (SRI), a prestigious organization known for pioneering scientific and technologi-cal advancements, from liquid crystals and solar energy to the ARPANET, the forerunner of the modern internet.

Engelbart envisioned a future where interactive computing would enhance human problem-solving and expand cognitive capabilities. He argued that computers could evolve from mere data-processing machines into tools that actively augment human intelligence, helping people struc-ture knowledge dynamically, analyze complex problems, and improve decision-making. It was a radical vision at a time when computers were programmed using punch cards and operator consoles.

In the mid-1970s, Engelbart created the Augmented Knowledge Work-shop, a revolutionary environment filled with workstations that featured multiple windows, integrated graphics, real-time collaboration tools, and a device he called a "mouse." His system, Augment, transformed how teams worked together, fostering a new way of thinking about digital collaboration.[3]

A brilliant yet soft-spoken visionary, Engelbart anticipated many of today's AI tools. He foresaw how machines could process complex ideas, recognize patterns, and assist humans in making better decisions. For de-cades, his vision of augmentation remained at the heart of computing.

The concept of "augmentation" of our capacity as humans is a pow-erful one for today. Now, we stand at the next great leap in augmenta-tion, where AI becomes a profoundly personal extension of the self and more. "This is the real magic of AI," says Harper Carroll, the founding AI engineer whose company was acquired by NVIDIA and is now a widely followed AI educator. "We're not just augmenting intelligence anymore—we're augmenting the human experience itself."[4] We're about to radically alter human capability and perhaps even what it means to be a human.

At the heart of this transformation is a fundamental shift in the nature of identity. Who are you? How do you see yourself? How do you present yourself to the world? Your identity is more than just a name, a passport, or a health insurance number. It's the sum of your characteristics, experi-ences, and actions.

You have a physical identity: your gender, race, appearance, voice, and how you move through the world. You may see yourself as a redhead, tall, tattooed, left-handed, or plus-sized. You also have a social identity, shaped by your roles and affiliations. You might be a parent, a student, a home-owner, a retiree, or a member of a professional organization. Your cultural identity includes your beliefs, preferences, and affiliations. You may identify as a Christian, atheist, feminist, social activist, Swiftie, or Lakers fan. Then there's your educational and professional identities; perhaps you're a teacher, an artist, a lawyer, or a labor union member.

But identity isn't just about how you see yourself, it's also about how others perceive you. Someone might describe you as brilliant, charismatic, a troublemaker, a slacker, or a leader. Your behavior expresses your identity in the physical world, and increasingly, digital technologies are capturing this behavior as data.

Every action you take—buying a dress at Zara, ordering a drink at a bar, passing an online course, posting on TikTok, attending a rally, swiping right on a dating app, listening to a playlist—creates a digital record. These actions form a digital mirror image of you, a lifetime's accumulation of data that reflects your identity.[5]

In the digital age, identity is no longer something you claim; it's something you and others are continuously shaping, tracking, and interpreting. AI is about to take this one step further. When you integrate AI into your digital identity, it metamorphoses from a passive cache into a dynamic thing, alive with intelligence. This digital version of you isn't a static profile; it's an agent acting on your behalf, making decisions, managing tasks, and learning from every interaction. That's why we've called this emerging category of artificial intelligence *identic AI*.[6]

## The Impact of Identic AI in Everyday Life

Identic AI is the most important of a new class of systems where AI agents are designed to operate autonomously. Your *identic agent* is a digital consigliere: a knowledgeable, intelligent, and capable extension of yourself.[7] It doesn't passively record your activities; it intimately knows you and your habits, preferences, and priorities. More than just an assistant, it's an evolving, intelligent partner that works proactively to enhance your life. More than your digital replica, it will become infinitely smarter, more knowledgeable and capable than you.[8] Your agent has a different "brain" than yours. As Eric Schmidt, former CEO of Google, provocatively asks: "What happens when each one of us has access, in our pocket, to intelligence equivalent to the smartest human available for every problem?"[9] In such a world, identic AI is more than a tool. It's a transformative force in how we think, decide, and act.

For individuals, an identic agent could act as a multilingual personal guide, manage daily schedules, curate media, facilitate relationships, and optimize productivity. It understands your emotional state, helps you set and achieve goals, and even nudges you toward healthier habits. Rather than creating an echo chamber for your views, it could be a devil's advocate that helps you think bigger and learn.

For professionals, an agent could redefine productivity, operating like a strategic adviser that anticipates challenges and initiates solutions. Unlike traditional AI copilots that assist with tasks only, an identic agent could negotiate contracts, initiate high-stakes business strategies, and collaborate with other AI systems. It would function as a partner, shaping outcomes with the foresight and adaptability of a seasoned executive.

Peter Diamandis, founder of the XPRIZE Foundation and co-founder of Singularity University, envisions this transformation. "Everything in my brain is capturable by my identic agent," he explains. "Like other managers, I have employees who do things for me—taking calls and making decisions—but they're imperfect. Once you have an identic agent you trust, you can replicate it infinitely, and the possibilities are extraordinary."[10]

Diamandis put this to the test. He built a digital representation of himself called "Peter Bot," training it with his books, interviews, and speaking engagements. "Peter Bot did a better job than I in representing my thoughts. It remembers everything I did. It speaks more clearly than I do. It's a smooth and compelling speaker." Rather than being threatened, he sees this as liberating his capacity to do so much more.

Andrew Ng, chairman of LandingAI, sees a future where everyone has an "army of assistants." "Human intelligence is expensive," he says. "If we can make intelligence cheap and give it to everyone, extraordinary things can happen."[11]

## The Double-Edged Sword of Identic AI

Yet, with immense potential comes equally profound risks. While many discussions about AI focus on job displacement, surveillance, disinformation, and autonomous military systems, identic AI presents unique concerns that go even deeper.

Who controls our personal AI agents? Will they truly serve us, or will they be owned and influenced by large tech companies and governments? What happens when your personal superpower, your identic agent, falls into the hands of others?

Beyond privacy and ownership, how do we integrate a superpower into our daily life? Managing one's physical self is hard enough. What happens when you must also manage a digital version of yourself?

What about inequality? If identic AI is a fundamental advantage, will we create a new class divide between those with digital superpowers and those without? Will society split into knowers and know nots, doers and do nots? This applies to companies and countries, too.

The deeper issue is agency itself. How much control will we truly have over a piece of AI that can act, decide, and shape our lives? Where do our choices end and the AI's decisions begin? What if our digital sidekick subtly shifts our perceptions of the world without our conscious knowledge? We wrestle with these profound questions in the pages ahead.

## A New Era of Human-AI Integration

Identic AI is more than another technological shift; it's a redefinition of human agency and the most important development in AI for each of us. It has the potential to empower us, helping us navigate an increasingly complex world, or to subjugate us, as it blurs the line between human and machine intelligence. We must ask: Are we the authors of our own lives, or is the digital extension of ourselves overwriting us?

This is the seismic shift we must control. Identic AI will reshape the future of society, industry, and what it means to be human. It will change our concept of identity itself. While its potential is revolutionary, it raises an even greater question: In this future, how do we humans remain in charge?

# WHY NOW? THE DRIVERS OF IDENTIC AI ADOPTION

Over the past decade, artificial intelligence has become the epicenter of technological innovation, attracting billions of dollars in investment from leading companies worldwide. Major players like OpenAI, NVIDIA, Google, Microsoft, and Meta have poured unprecedented resources into AI research and development, competing to shape the future of digital intelligence. In 2025 alone, the global AI market was valued at approximately $200 billion, with projections suggesting it could reach over $1.8 trillion by 2032.[12] Leading US technology companies, including Amazon, Microsoft, Alphabet (Google), and Meta, collectively invested over $300 billion in AI infrastructure in 2025.[13] And that's just the USA! There are also massive investments underway in China, across Europe, and in many other countries.

The capabilities of these systems have expanded at an exponential pace. Just five years ago, language models could barely hold a conversation.

Today, AI systems can generate essays, write software, analyze images, and even diagnose medical conditions with a degree of accuracy once thought possible only of human experts. Advances in AI hardware have helped make this leap possible, including NVIDIA's designing GPUs for high-powered AI computations. These powerful processors enable the training of massive AI models on billions of data points in just weeks or even days, a process that once required months. Projections into the near future suggest we're on the cusp of another leap: AI systems with billions of parameters could soon reach capabilities to perform specialized tasks across multiple domains simultaneously.

If current trends hold, it's likely that by 2030, or perhaps much sooner, AI systems will possess a level of contextual awareness and adaptability that allows them to operate seamlessly in almost any human environment, pushing the boundaries of what we thought possible. Often referred to as *artificial general intelligence*, or AGI, the timeline on which AI will surpass human intelligence remains open to much debate and speculation. Dario Amodei, co-founder and CEO of Anthropic, expects "human-level" AI within two to three years.[14] Sam Altman, CEO of OpenAI, believes we will reach AGI as early as the end of 2025.[15] Says Peter Diamandis, "By 2029, AI will be as intelligent as the entire human race. And soon enough, a billion-fold more intelligent than us."[16]

This surge in capability, powered by intensive investment and technological innovation, lays the groundwork for a new generation of autonomous, proactive systems that collaborate, anticipate, and adapt to users' evolving needs. We dive into the breathtaking technological foundations of identic AI in chapter 3.

But technological progress doesn't happen in isolation. Identic AI is emerging amid global crises that need novel interventions. From growing urban dysfunction and crumbling infrastructure to escalating healthcare costs, the needs of an aging population, and an ever-warming planet, our systems are under increasing strain. Identic AI is a direct response to these pressures, offering hyperlocal solutions that would have been inconceivable even a decade ago.

## Economic Pressures and the Demand for Greater Productivity

Heightened competition, resource scarcity, and increased volatility characterize the global economy. Businesses face complex challenges: fluctuating markets, supply chain disruptions, and relentless demands for productivity. These pressures have intensified the need for new efficiencies. Yet traditional methods of streamlining operations are reaching their limits. Humans can handle only so much data and complexity before productivity gains plateau.

Enter AI agents. Unlike conventional AI that reacts based on pre-set rules, identic agents can have the context-awareness and decision-making abilities to manage complex processes proactively. In manufacturing, for instance, autonomous AI systems can predict demand shifts and respond by adjusting production schedules, ordering materials, and negotiating with suppliers, all without human intervention. The business is adapting itself in real time, something no human team could manage at scale.

For individual professionals, identic agents will take on repetitive, data-heavy tasks, acting as collaborative partners that streamline workflow and enhance decision-making. On a recent earnings call, Alphabet Inc. CEO Sundar Pichai said that AI agents now generate more than 25 percent of all new code at Google.[17] While engineers still review and accept the AI-generated code, Pichai said the result is a reimagined workforce where engineers can do more and move faster than before.

In 2025, many companies released suites of tools to accelerate agent-based applications, that include agent development kits for building autonomous systems.[18] Together, these tools mark a shift toward software environments where AI agents and users work in tandem toward shared goals, ushering in a new era of cooperative intelligence.

As businesses embrace this model, economies are witnessing the beginnings of a significant shift: productivity gains driven by human-AI partnerships. As more firms rely on identic AI to stay competitive, a new economic landscape is emerging: one that depends on autonomous systems to maintain stability and growth in increasingly turbulent times. Productivity is just the beginning.

## The Evolving Work Model: From Remote to Digital Partnerships

The global pandemic of the early 2020s ushered in widespread acceptance of remote and hybrid work models, proving that people could collaborate productively outside traditional office spaces. While the shift responded to necessity, millions came to prefer the model worldwide. However, the model introduced coordination and burnout challenges and the need for adaptive, self-directed work environments.

As the traditional office dissolves, the concept of *workplace* has expanded, creating new demands on technology to support hybrid work, with distributed teams. With its context-aware intelligence, identic AI fits naturally in these evolving work models. Our digital agents will bridge the gap between distributed human teams, enhancing focus, providing strategic insights, and handling coordination beyond simple scheduling or task management.

The rise of AI-driven collaboration does not only change *how* we work, but also redefines *who* and *what* is doing the work. As identic agents become central to professional productivity, businesses will face a new set of questions. What skills will matter in an AI-augmented workforce? How will companies evaluate employees when individual performance inextricably links to AI capabilities? How do we balance corporate data ownership with personal digital autonomy? These challenges will reshape employment and compensation models and even redefine human labor.

Traditionally, companies have evaluated experience, education, motivation, and intelligence when hiring employees. In the future, they will be just as interested in how well a candidate has developed and utilizes their personal work copilot. If AI agents like Peter Diamandis's Peter Bot can take on almost infinite tasks, what kind of variable compensation models will make sense where AI supercharges productivity?

What happens when an employee leaves the company? Do they take their agent, along with its deep knowledge of the firm, with them? Some proprietary information will surely remain corporate intellectual property, but at the same time, an employer cannot claim ownership over an individual's digital identity. And what about intellectual property rights? If a report or book is codeveloped by a human and their identic agent, who owns the rights? Who holds the copyright?

Beyond these corporate challenges, AI agents raise broader concerns about labor markets. Carl Benedikt Frey and Michael Osborne, in their study "The Future of Employment," estimate that "47% of total US employment is at risk."[19] Some industry players take an even more radical view, suggesting that most jobs will be eliminated, and that employment might become optional, more like a hobby than a necessity.[20] Yet, these techno-optimists offer few details on how displaced workers will earn an income in a post-work world.

## The Education Gap and the Rise of Private Tutors

Education systems worldwide are struggling to keep pace with the needs of digitally native learners. Traditional one-size-fits-all models are increasingly obsolete, unable to accommodate different learning speeds, interests, and skill sets. Today's students require personalized, adaptive learning experiences, approaches that cater to individual aptitudes and continuously evolve alongside their needs. Beyond foundational education, many individuals will need to reinvent their knowledge base multiple times throughout their careers.

With identic AI, education can finally move beyond the rigid, industrial-age model toward a more interactive, customized, and responsive system. AI-driven learning companions have the potential to transform education by tailoring curricula, generating real-time feedback loops, and facilitating peer-to-peer collaboration. These companions can identify a learner's strengths, detect areas that need improvement, and adjust lesson plans dynamically. For students with unique needs or learning challenges, identic AI can create highly personalized educational experiences, increasing engagement and improving outcomes.

However, rethinking pedagogy is no simple task. We must distinguish between the mastery of knowledge—where there are clear right and wrong answers—and the broader dimensions of learning, such as critical thinking, problem-solving, research, collaboration, and the ability to see the big picture. Identic AI can assist in all of these areas, but its role must be carefully considered.

As students transition into the workforce, their identic agents can act as career-long learning facilitators, helping them acquire new skills and stay ahead of economic and technological changes. These AI-powered tutors will not only help individuals master content; they will recommend discussion groups, courses, and learning strategies fitted to personal and professional growth. Instead of curating existing knowledge, identic AI can actively expand users' intellectual horizons so that they can adapt to job markets that demand continuous learning.

But this shift presents difficult questions. In the past, individuals made conscious or unconscious choices about what knowledge to retain and what to store externally, offloading facts and data to the web, where they are accessible through search engines. Now, the challenge is exponentially greater when your identic agent possesses all recorded information and can retrieve, analyze, and synthesize it instantly. What should we prioritize learning ourselves, and what should we delegate to AI? And beyond mere knowledge storage, identic agents will also have the capacity to think. They will be able to solve problems, make recommendations, and even take actions on our behalf. This raises an even deeper question: What aspects of cognition will we outsource to our AI superpowers? As we increasingly rely on these digital extensions of ourselves, how will this reshape human learning, decision-making, and autonomy?

## Soaring Healthcare Costs and the Promise of Preventive, AI-Driven Care

Healthcare systems worldwide are straining under rising costs, limited resources, and growing demand. As chronic conditions increase and populations age, the gap between healthcare supply and demand becomes unsustainable. The traditional model, where access to care depends on the availability of doctors, nurses, and medical facilities, is no longer working. Today, most systems manage the imbalance between demand and supply through rationing. Most developed countries with single-payer healthcare systems allocate care according to medical urgency. In the United States, market forces often dictate access; the ability to pay determines priority. Either way, demand far exceeds supply, and training and hiring enough medical professionals to meet the growing need is unfeasible.

Identic AI can alleviate these constraints. AI can empower healthcare providers with superhuman diagnostic abilities, reduce errors, accelerate treatment, and cut costs. More importantly, it shifts care from reactive treatment to preventive and continuous health management. Personal AI health companions can monitor daily health metrics, detect early signs of illness, and recommend preventive measures such as dietary adjustments, stress management, or timely medical checkups. These AI-driven systems can also coordinate care, ensure medication adherence, and prompt individuals to seek professional attention before minor issues escalate into medical crises.

For older adults, AI companions can provide cognitive support, assist with daily tasks, and offer social interaction, mitigating loneliness while enabling independent living. These systems don't just serve as reminders for medication or routine checkups; they act as proactive caregivers, offering reassurance and companionship.

Beyond individual care, identic AI's autonomous multiagent systems could revolutionize healthcare logistics, optimizing resource allocation, managing patient flows, and coordinating treatment across facilities to ensure seamless care. This would alleviate the administrative burden on healthcare professionals, allowing them to focus more on patient outcomes.

Identic AI won't replace human caregivers, but it could enhance their capabilities, making high-quality healthcare more accessible and sustainable. The question now is not whether AI will transform healthcare, but how we will shape this transformation so that it benefits everyone.

## Government and Public Policy:
## Meeting Complex Needs with Adaptive Intelligence

Government institutions are under pressure to address complex issues like housing, migration, infrastructure, and climate change while operating within tight budgets and mounting debts. Bureaucratic inefficiencies, limited transparency, and slow response times are no longer acceptable in a fast-moving world. Citizens expect effective governance, and AI is becoming a powerful tool to achieve this.

Take immigration, a highly contentious issue in the 2024 elections in the United States. Managing migration effectively involves coordinating efforts across border security, housing, employment, education, and healthcare—all complex, data intensive, and highly dynamic. Identic AI can be a game-changer, as bureaucrats tap autonomous AI systems to track migration patterns, analyze workforce demands, and predict housing and infrastructure needs. By continuously analyzing migration data, detecting trends, and making adaptive recommendations, identic AI could help policymakers manage immigration flows more effectively and support the integration of newcomers with investments in training, housing, and infrastructure. In other words, AI becomes more than a governance tool; it acts as a responsive partner in building resilient communities that balance social needs with economic opportunities.

Identic AI's potential extends beyond administrative efficiency: It can revolutionize how citizens engage with government. Today, people participate in politics passively or only during voting cycles, yet the most pressing societal challenges require continuous civic engagement. An identic agent can act as a civic adviser, helping us understand legislation, track local developments, and advocate for our interests. Citizens could have AI-powered policy assistants that analyze government proposals, generate informed arguments, and even draft public comments or petitions. Identic AI could also give people real-time insights into how policies affect their daily lives, empowering them to participate in decision-making.

However, the same AI tools that can personalize governance could also reinforce digital echo chambers, exacerbating polarization. If AI-driven political advisers filter information too narrowly, citizens may engage only with perspectives that confirm their existing biases, deepening divisions rather than fostering informed debate. Governments must also contend with the risk of AI-generated misinformation shaping public opinion. The challenge will not just be deploying AI in governance but ensuring it strengthens democratic participation rather than undermining it.

## Social Fragmentation: Rebuilding Connection and Community

While the digital revolution has made communication more accessible than ever, paradoxically, it has also contributed to social fragmentation and isolation. In an age where people connect primarily through screens and virtual platforms, many report feeling more disconnected, struggling to find genuine human interaction, and navigating increasingly polarized social spaces. Compounded by remote work, geographic mobility, and polarized media, this social fragmentation has heightened the need for new forms of connection and support.

Indeed, loneliness and social isolation have significant social costs, impacting both individual well-being and broader social health. Research indicates that individuals experiencing loneliness are at higher risk for various health issues, including heart disease, stroke, type 2 diabetes, depression, anxiety, and dementia. These health complications affect individuals and impose economic burdens on healthcare systems.[21]

Economically, the repercussions are substantial. In the United States, issues associated with chronic loneliness, such as depression, suicide, and addiction, cost the economy approximately $960 billion annually.[22]

Identic AI has the potential to address these issues by fostering community and providing companionship. Through AI-driven personalization, your identic agent could help you form deeper connections, whether by identifying compatible partners, strengthening friendships, or supporting emotional well-being. For instance, it could engage in preliminary interactions with a prospective partner's agent to determine compatibility before you even meet.

In therapeutic settings, AI companions could support mental health professionals by offering real-time mood tracking, facilitating guided cognitive behavioral therapy exercises, or providing immediate coping strategies between therapy sessions. They could also offer supportive dialogue, recommend social activities tailored to an individual's emotional needs, or connect users with peer-support networks and professional help when necessary.

Beyond individual companionship, identic AI can act as a bridge to larger communities, connecting people based on shared interests, goals, and values. By analyzing user preferences and interaction patterns, identic AI can introduce individuals to like-minded people, forming virtual communities where users can collaborate, learn, and support one another.

While AI companions can provide comfort, connection, and even a sense of intimacy, there is a darker side to forming deep emotional bonds with artificial entities. Millions of users on platforms like Replika and Character.ai have built relationships with AI companions, with some even considering them romantic partners. However, unlike human relationships, AI companions do not have genuine emotions, independent agency, or reciprocal attachment; they simulate affection based on algorithms designed to maximize engagement. Moreover, these relationships exist within platforms controlled by corporations that can alter or remove AI personas at any time, as seen when Replika's developers unexpectedly stripped its chatbots of romantic capabilities, leaving many users devastated.

This raises ethical concerns about emotional dependency, manipulation, and the potential for users to prioritize AI interactions over real human connections, further deepening isolation rather than alleviating it. As AI-driven companionship becomes sophisticated, the question remains: Are we genuinely enhancing human connection, or are we replacing it?

## THE INVISIBLE FORCE RESHAPING HUMANITY

As identic AI emerges to meet urgent societal demands, it offers more than just solutions to today's pressures: It opens pathways for profound transformation. Eric Schmidt recently observed that "Society currently has no framework, no language to describe the transformative impact of intelligence at this unprecedented scale." He calls the looming transition "under-hyped" because "people can't fully grasp what's about to arrive."[23] In a few short years, this new wave of AI will reshape how we work, learn, govern, and live, not only addressing inefficiencies but also unlocking new opportunities for connection, creativity, and resilience on a global scale.

Identic AI is a force of immense power, whether for good or bad. It's arriving fast, and for most, it's an invisible, stealth presence creeping into our lives.

Singer-songwriter Bonnie Raitt's 1994 hit reassures, "Don't worry, baby, that's just love sneakin' up on you."[24] But has there ever been a time in modern history when a major innovation sparked such a mix of excitement, unease, or outright fear?

Is this moment like The Seekers singing, "There's a new world com-
ing, and it's called the Promised Land,"[25] or Frank Sinatra crooning, "The
Best Is Yet to Come"?[26] Or is it more like the stealth shark in *Jaws*, Anton
Chigurh in *No Country for Old Men*, or the weird sisters in Shakespeare's
*Macbeth*, who warn, "Something wicked this way comes"?

This book explains that it's neither . . . or both. Clearly, we are in awe of
the emerging potential to transform humans and our world for the better,
but we are not technology determinists. Rather than forecasting a future,
we need to choose it and achieve it, each of us and together. Our goal is to
contribute to a promise fulfilled and a peril unrequited.

In chapter 2, we explore how the very concept of identity is changing.
Our identities are becoming digital, enriched with data, endowed with
intelligence, and increasingly under our control. Then, in chapter 3, we
raise a fundamental problem: the concentration of these AI capabilities
in the hands of a few dominant tech companies. We make a case for a
self-sovereign digital identity and outline the structural shifts required in
AI development, ownership, and the broader technology industry to support this vision.

Having examined the technological underpinnings of identic AI, we
explore its impact across six core domains, each in a dedicated chapter,
to illustrate how this technology will shape our daily lives and redefine
society.

Chapter 4 explains how identic AI will revolutionize our professional
lives, supercharging productivity and transforming workplace dynamics.
AI agents won't just assist with tasks; they'll collaborate, manage workflows, and even initiate projects, fundamentally altering how we think
about employment and career progression.

Chapter 5 explores education and lifelong learning. It demonstrates
how identic agents can offer personalized learning paths and on-demand
skill development, supporting learners of all ages and empowering them
to adapt to an evolving economy.

Moving into the creative domains, chapter 6 examines artistic creation and media consumption, revealing how identic AI will influence what we see, hear, and create. While AI unlocks unprecedented creative potential, it also raises critical questions about authorship, originality, and influence. We also explore how culture needs a new business model and how this technology could hurt or help artists reap rewards for their creative endeavors.

Chapter 7 turns to urban life and local democracy, where identic AI will revolutionize how humans live in cities and how cities function, from optimizing infrastructure to enhancing public services and citizen participation. We argue that identic AI could facilitate a new kind of "cognitive city," where AI augments democratic engagement and helps manage complex societal challenges.

Chapter 8 shows how identic AI will shift the prevailing healthcare paradigm from reactive treatment to preventive care. These systems will monitor wellness, predict health issues, coordinate medical interventions, and accelerate biomedical research. But with AI playing an active role in personal health, where do we draw the line between assistance and control?

Chapter 9 delves into interpersonal relationships, community, and belonging, highlighting how AI companions could serve as confidants, social facilitators, and even a source of companionship. But does AI strengthen social bonds, or replace them? This chapter examines the promise and potential perils of AI-mediated relationships, ultimately providing a holistic view of how this powerful technology stands to redefine society and our sense of self.

The final chapters provide essential frameworks and practical guidance for navigating this transformation responsibly.

Chapter 10 outlines seven core principles—reliability, transparency, agency, adaptability, fairness, accountability, and safety—that will shape how we design, deploy, and govern identic AI to ensure it serves humanity. These principles provide a foundation to guide individuals, businesses, and policymakers as they steer the development and use of identic AI.

Chapter 11 shifts focus from broad principles to practical guidance for those facing the direct impacts of AI-driven change. It offers insights for individuals and organizations on how to adapt to new realities in work, education, media, urban life, healthcare, and community engagement. It encourages readers to view these changes not as disruptions to endure but as opportunities to thrive, emphasizing proactive engagement and a commitment to continuous learning as we move into an era of unprecedented innovation and interconnectedness.

These final chapters serve as both a compass and a tool kit to help readers navigate the AI-powered future with clarity, adaptability, and purpose. After all, with the startling new possibilities we describe come profound shifts, as well as a potential dark side. As identic AI becomes deeply embedded in daily life, it will shape our choices, habits, and beliefs in ways we may not fully recognize.

This influence is not theoretical. It's already here. AI systems subtly shape our behaviors, filtering the news we see, prioritizing career suggestions, nudging us toward particular health choices, and even influencing our social interactions. Over time, these recommendations become seamlessly integrated into our routines, guiding us in ways that feel natural but are, in fact, engineered. A personal AI companion might encourage wellness habits, suggest specific films, or facilitate professional networking, all under the guise of personalization. Yet, in doing so, it becomes an invisible architect of our lives, subtly steering us toward particular trajectories designed by opaque and unknown forces.

This shaping force extends beyond the individual. In workplaces, identic AI doesn't just assist employees: It influences team dynamics, assigns tasks, and determines whose insights get amplified in meetings. These microdecisions, driven by algorithmic optimization, redefine workflows, leadership structures, and corporate culture over time. AI doesn't simply support work; it gradually reconfigures it, embedding strategic priorities that align with an organization's goals rather than individual agency.

The same applies at a societal level. AI-driven content feeds, digital assistants, and media algorithms continuously refine and reinforce perspectives, creating feedback loops that shape public discourse, political affiliations, and cultural values. This invisible guidance is not inherently good or bad, but it is powerful and often unaccountable. What we consume, believe, and prioritize is increasingly curated, sometimes reinforcing our worldviews and other times subtly shifting them.

As identic AI assumes more complex roles, profound ethical questions emerge. How much decision-making should we delegate to identic agents? If an AI adviser recommends a medical treatment or an investment strategy, it may act in a person's best interest. But who defines that "best interest," and according to what criteria? AI companions, designed to be helpful and intuitive, could become trusted confidants, blurring the line between assistance and influence.

With increasing reliance on identic AI, a deeper dependency could form. What happens when we outsource thinking, decision-making, or even relationships to AI? If a digital agent anticipates our needs before we recognize them, what role does human intuition play? Could a generation raised alongside AI companions find it more natural to confide in digital entities than in human friends? While these systems provide convenience and tailored support, they also risk eroding self-reliance, prompting difficult questions about where human agency ends and AI influence begins.

Perhaps most consequentially, identic AI challenges the future of human work and identity. As AI systems perform both analytical and creative tasks, the nature of employment may shift from a "human-in-the-loop" or "human + machine" scenario to outright displacement. What happens to a person's sense of purpose when agents take over tasks traditionally associated with distinctly human traits and capabilities? In healthcare, education, and customer service, AI could assume responsibilities requiring empathy, problem-solving, and social connection. This shift may fundamentally alter career paths, forcing society to redefine purpose and fulfillment in a world where traditional jobs are no longer a given.

The urgency and gravity of this transformation cannot be overstated. With each task delegated to our agents, with each choice informed by

algorithmic suggestions, we unknowingly participate in a silent revolution. We are placing increasing trust in an intelligent system that is both
immensely powerful and largely opaque. This system can enhance optimize, and connect, but it can also persuade, filter, and redirect. Identic AI
is not just another convenience; it is becoming a partner in decision-making, an architect of experience, and an invisible force shaping modern society in ways that demand critical attention.

In the chapters ahead, we explore how this profound shift plays out
across work, education, healthcare, relationships, and governance, uncov-
ering the extraordinary potential and peril of identic AI. This journey will
reveal the dual nature of this transformative force, illuminating both its
promise and the intricate web of influence it weaves, often hidden in plain
sight.

## Chapter 2

# THE EVOLUTION OF IDENTITY IN THE DIGITAL AGE

Throughout history, each new form of media has expanded human reach, dissolving barriers of time, space, and even mortality. This extraordinary capability continually revives fundamental questions of identity: What defines us? What does it mean to be human? How do we perceive ourselves and project that identity to the world, both physically and digitally? As Marshall McLuhan noted, over time, the medium shapes the message.[1] In turn, people shape media, and media reshapes people. Our minds evolve, our institutions adapt, and society transforms in response. The human experience changes. We work with digital tools and live in connected, monitored, and recorded environments.

Today, we each have a sense of our self, a sameness of our essential character in different contexts or media that we call an identity. In the digital age, that identity has a mirror image, not in the sense of a looking glass, where we see the inverse of what others see, but in the sense of computer disk imaging, a copy of ourselves as expressed in data ranging from our DNA and our relationships to our every action and transaction online.

The data reveals who we are and how we live our lives: our beliefs, personality traits, gender, race, appearance, athleticism, culture, medical history, vocation, avocations, intelligence, level of education, memberships, political views, purchases, finances, travels, hobbies, and lots more. Think about what your social media, search engines, email, credit card, telephone, and streaming service providers know about you, the people in your life, your whereabouts, and your daily routines. Your blob of data makes you truly unique, and the data visible to the world shapes how the world sees you.

We all create this data, but none of us owns or controls it. We don't even know how much personal data exists, let alone what it is, where it resides, and who's doing what with it. Why? Because we live our lives across other people's data silos. It begins in the womb, with that first sonogram our mother's obstetrician captures, and ends with the tombstone marking our final resting place, and the last bits of our data in the undertaker's cloud.

With new technologies, we can now recover this data so that we can manage it responsibly for our benefit. Add in the power of AI, and we can use our data to train AI agents that represent us in the world. With AI, your digital persona is not just a blob of data. It is your digital self; it can read your life's data and recognize the patterns. This digital self is intelligent and ready to help, perhaps with its own agency.

In this chapter, we explore how identity is evolving in the digital age, tracing its journey from traditional notions of selfhood to the emerging concept of the digital self, shaped by identic AI. We begin by examining the history of identity—its philosophical roots, psychological development, and sociological dimensions—and how these ideas have adapted in response to technological advances. Today, the data that defines our identities is controlled mainly by corporations and governments and used in ways that often undermine individual autonomy and privacy. We critique what Don has called "digital feudalism" and call for a shift in the paradigm of data in the digital economy, one where individuals own and control their digital identities, ensuring they reflect their true selves while maintaining privacy and security.[2]

The discussion of digital feudalism sets the scene for a three-stage evolution in the digital self. Stage I addresses the problem of identification, advocating for self-sovereign systems that allow individuals to control their identifiers without relying on centralized authorities. Stage II expands this to the concept of a digital identity: a comprehensive, user-controlled repository that consolidates medical, financial, and personal data into a secure digital wallet. Stage III envisions the rise of identic AI, where static digital identities evolve into intelligent, autonomous companions capable of acting on our behalf. These agents have the potential to redefine how we interact with the world, from improving our decision-making to representing us in both personal and professional contexts. However, this shift also raises profound questions about control, authenticity, and the boundaries of identity in a world where humans and their digital selves coexist.

These issues of digital identity and the digital self are central to understanding the transition to identic AI. A comprehensive digital identity enables identic AI to learn from our personal data, aligning with our values, preferences, and goals. In doing so, it provides the foundation of how these systems operate and represent us, transforming static information into dynamic intelligence and making our digital selves capable of meaningful action. By owning and managing these identities, we maintain agency over how we are represented and ensure that AI complements rather than compromises our humanity. This evolution not only challenges traditional notions of identity but also demands that we rethink our relationship with technology, ensuring it enhances our lives while reflecting our core human values.

# A BRIEF HISTORY OF IDENTITY

To begin our journey on the transformation of identity by new technologies, we need to understand the concept and its history. Identity begins to form in childhood as we develop our self-concept throughout life.

Familial, social, and cultural rituals and interactions, along with how others perceive us, all shape our identities. Over our lives, we accumulate identifiers, the essential artifacts of identity. They originate mainly from governments and corporations: our birth certificate, social security number, driver's license, diploma, banking and credit accounts, frequent flier membership, and marriage license, to name a few. These identifiers hold meaning both personally and socially, shaping our relationships with key institutions and enabling our participation in different spheres of society.

The word *identity* comes from the Latin *identitas*, which encompasses our self-image and sense of similarity to others. Identity serves several purposes. It acts as a framework for self-regulation, providing meaning, direction, and a sense of control. It helps maintain internal coherence and guides behavior so that we can plan and set objectives. As an active and ongoing process, identity significantly affects how we adapt to life and achieve well-being. However, aspects of identity often stem from characteristics beyond personal control, such as our genes and family—nature and nurture.

The concept of identity has evolved from rich cultural, spiritual, and philosophical traditions. Eastern philosophies such as Taoism emphasize the fluidity of identity and the interconnectedness of all things. Hinduism views individual identity (Atman, the true self) as an expression of a larger cosmic reality (Brahman, the universal spirit). Buddhism challenges the permanence of identity, proposing that the self is a construct shaped by transient experiences and perceptions (anatta, the nonself).

Although the ancient Greeks had no fully developed theory of identity, they laid the groundwork for understanding our sense of self. In Plato's dialogue *Theaetetus* (369 BCE), we find the earliest "law of identity." Often summarized as A = A, this law asserts that a thing is identical to itself and distinct from every other thing. It is a simple and profound equation that expresses differentiation.

The law of identity became a cornerstone of Western philosophy for generations of thinkers, from Aristotle to Wehmeier.[3] Early discussions often centered on objects rather than people. Philosophers in classical Greece grappled with enduring questions: What are the essential properties of an object's identity? Is an object the sum of its parts, or does it have an underlying "essence" that persists amid change? Is there a threshold of change beyond which identity is lost?

Here's a classic thought experiment known as the Ship of Theseus from Plutarch.[4] Over time, the Athenians replaced the rotting timber of the ship with new planks. Is it still the same ship? More than a thousand years later, Thomas Hobbes added a twist: What if the Athenians created a second ship from all the decayed timber?[5] Which of the two ships should we consider the original ship of Theseus: the one with fresh planks or the one from recycled timber? What are the criteria for answering such a question about something or someone?

In the seventeenth century, John Locke shifted the discussion by focusing on personal identity. Rather than defining identity in terms of one's physical body or spiritual essence, Locke proposed that individual identity is rooted in consciousness, particularly memory. Although our bodies change over time, our ability to remember and reflect on our experiences defines us. Physical continuity was less significant than a continuous sense of self-awareness. In other words, identity resides in our mind—our conscious experience.

David Hume took a skeptical view of this notion of a "continuous self." He argued that what we think of as a "self" is a bundle of fleeting perceptions and experiences bound together by habit rather than by any underlying essence. He viewed identity as fluid: an aggregate of moments, not an enduring whole. Hume opened the door to a more dynamic understanding of personal identity.

As people moved through the Enlightenment to the nineteenth century, their concept of identity expanded beyond the individual. With the rise of industrialization and urbanization, societies became increasingly interconnected. Philosophers recognized that both internal and external forces shape identity. Sociologists examined how social roles and group affiliations influenced people's sense of self, and how a person's identity relates to collective identities, such as family, nationality, professions, or social class. Under this view, identity was not just personal memory or self-reflection but also our positioning in society.

In the twentieth century, psychologists like Erik Erikson viewed identity as a continuous process rather than a static characteristic. Erikson introduced the idea of "ego identity," which he saw as an evolving sense of self that took shape as individuals navigated life's challenges and interacted with society.[6] He recognized adolescence as a particularly intense period of identity formation: a time marked by experimentation, self-discovery, and sometimes crisis. Building on Erikson's ideas, psychologist James Marcia developed a framework for understanding different identity statuses in terms of exploration and commitment.[7] By studying how individuals chose and committed to different identities, Marcia expanded our thinking of identity as a dynamic process rather than a fixed trait.

This new understanding—that identity was a product of inner consciousness and external forces—led to the idea that people could "perform" or deliberately express their identities in certain social contexts. Sociologists like Peter Burke noted that identity could be something we project outward, shaping how others perceive us and influencing how we act in specific social settings.[8] In this view, identity was not only something we possessed; it was something we did. We projected it onto the world. In 1959, the sociologist Erving Goffman used the metaphor of theater, where actors perform their characters on a stage through their verbal and physical actions and reactions, their silence and stillness, navigating settings constructed by each other and the audience's perceptions.[9] Striving to cohere and adjust to various social contexts, actors interact with others who are also attempting to present themselves in a specific light.

The digital age has expanded how we present ourselves to the world. With the rapid adoption of digital technology in the late twentieth century (in particular, the World Wide Web), individuals had an entirely new stage for their projections or even reinventions of their identities in previously impossible ways. Through social media and online avatars, we could experiment with new personae, creating multiple versions of self-expression in the digital realm. Each of us had to think about how we wanted to present ourselves to a bigger and different world. Researchers like danah boyd and Knut Lundby observed that this ability to craft digital identities freed individuals from the constraints of their physical bodies and societal expectations, opening new avenues for self-exploration and social connection.[10] It also came with a world of problems, as we shall see.

Let's summarize the pertinent ideas so far. Identity is about things *and* people. Each of us has an identity that is a combination of characteristics. Identity consists of information about us but much more: It includes our memory, self-awareness, consciousness, and intelligence that act on that data. Identity is not static but a continuous process that changes over time. We each have a self: a concept of who we are that we present to the world. The digital age opened new venues for us to present and represent that self. We'll come back to all this in a minute.

## "SERFING" THE INTERNET: DIGITAL FEUDALISM

Our daily experience in the digital age generates enough data for banks and online companies to profile us and sell that profile of our "virtual selves" to other entities. In their book, *Who Knows? Safeguarding Your Privacy in a Networked World*, Ann Cavoukian and Don Tapscott explained how valuable our data stream is to companies and governments, so much so that we now call data "the new oil."[11]

Now, as you spend hours online each day, this "virtual you" knows more than you do in many areas of your life. Unless your memory works differently from ours, you don't recall your exact location a year ago, what you bought, how much money you spent and received that day, what you posted online, what your heart rate was, which medication you took, or what diagnosis you received.

We create this data, but we don't have access to most of it, and we don't own it. Tapscott called it "digital feudalism," drawing parallels to the power dynamics of the medieval era. In feudal societies, laborers toiled on land they didn't own, generating value while landlords claimed the majority of the rewards. Today, data has become the defining asset. It is sown by users but harvested by digital landlords, including social media platforms, search engines, online marketplaces, financial institutions, and governments. What was once "surfing the internet" has become "serfing the internet," as Tapscott put it, with individuals unwittingly surrendering deeply personal insights in exchange for access to digital spaces controlled by corporate gatekeepers.

Yet even in this data-driven economy, the picture remains incomplete. Not everything we do generates data, and many of our actions still go unnoticed by the digital systems around us. The "digital version of you" is necessarily partial. And even when our behavior is recorded, it doesn't tell the whole story. Beneath the surface lie your private thoughts, fears, dreams, and desires—intangible elements of identity that are rarely captured and never fully understood by algorithms.

This disconnect reveals a deeper problem. The growing asymmetry is more than just an economic issue. It strikes at the heart of personal identity in the digital age. The information we generate isn't just transactional metadata; it is an intimate reflection of who we are. Yet, instead of demanding ownership of this digital self, we continue to occupy spaces where others dictate the terms, forfeiting control over our most valuable asset—our own data. Yes, you get lots of free services this way, from search to social media. So, why should you be concerned? Consider the trade-offs of this new form of feudalism.

First, as noted earlier, your data resides across separate databases of myriad entities. These databases are of two types: those that have been hacked and those that have yet to be hacked. Unless you're a hacker, you can't access your own data.[12] You can't use it to plan your life—your health, diet, budget, education, use of carbon, investment in solar panels or electric cars, retirement, and so forth. You can't aggregate and anonymize it for social purposes. As the pandemic showed, societies have no access to critical data in times of crisis.

Second, you can't reap rewards from your own digital assets, not for yourself, your family, or your heirs. While corporations and institutions profit from your data, you're left out of the value chain. Worse, you bear most of the risk: If a company misuses, loses, or exploits your data, it's you—not them—who deals with the fallout.[13] After the Cambridge Analytica scandal involving 50 million accounts, Facebook managed to expose nearly *420 million* user IDs and telephone numbers but didn't have to deal with the tsunami of robocalls and fraud alerts.[14] This is data malpractice.

Third, these entities are undermining your privacy. Like many of us, you've made the Faustian deal with technology companies, exchanging *your* data for *their* capability. Some say that privacy is dead and that those of us with nothing to hide should just "get over it."[15] We disagree: Your privacy is worth protecting. It contributes to your identity and your sense of self.

What's more, a government knowing everything about you is dangerous. Let's not assume that governments are benevolent. Often, they are not. Today, oppressive regimes use mass surveillance to control their populations. Corporations, too. What if a company had perfect information about you? Yes, it could tailor products and services. But it could also manipulate you in ways that weren't in your best interest. Remember that George Orwell's *1984* was a novel, not a playbook. Privacy is the foundation of freedom.

Further, our data is in myriad private silos, so it can't be anonymized and aggregated for social purposes. As the pandemic shows, there is no way for societies to access these data in times of crisis. So, for example, epidemiologists can't get access to critical aggregate data to identify a pandemic as it gets started or fight it once it spreads around the world.

# A THREE-STAGE APPROACH TO CREATING THE DIGITAL SELF

Regulators have proposed various ways to chip away at these problems, such as breaking up Amazon, Facebook, and Google as data monopolies or relying upon the European Union's general data protection regulation (GDPR). But they're woefully insufficient.[16] The GDPR is a partial measure and hypocritical in light of the EU common identity repository.[17] Rather, we need to fundamentally overhaul the legal frameworks that define data ownership along with the entire architecture of AI itself. This will ensure individuals, not corporations, have control over their personal information. In the digital age, protecting identity requires new mechanisms for establishing, managing, and securing our digital presence, shifting power away from centralized entities and toward the individuals who generate the data. To be sure, this is an ambitious undertaking but, as you will read, one that is achievable.

In the framework that follows, we outline a three-stage approach to creating the digital self. The framework describes a transformative journey from essential online identification to fully intelligent and autonomous digital companions that act as extensions of ourselves. Stage I focuses on solving the problem of identification, emphasizing the need for a self-sovereign system that allows individuals to own and control their digital identifiers without dependence on governments or corporations. By leveraging technologies like blockchain—a decentralized and tamper-proof digital ledger—tools like Civic Pass already demonstrate how digital identities can enhance privacy and security while facilitating trust in digital interactions. Stage II extends this concept to a full digital identity, consolidating diverse data (such as medical records, financial credentials, and social personas) into a single, user-managed digital wallet. This stage empowers individuals to selectively share or monetize their data and navigate the digital world with privacy and control. Finally, Stage III introduces identic AI, where digital selves evolve from static data repositories into intelligent, autonomous companions. These virtual selves, imbued with our values and capable of independent action, have the potential to assist in decision-making, advocate on our behalf, and redefine how we interact with society, raising profound questions about identity, agency, and personhood in a world where humans and AI coexist.

## Stage I: Solving the Problem of *Identification*

Let's start with the thorny problem of identification—the foundation of identity in the digital age. How do you prove who you are online? How do you know who anyone else really is? How can you share only the necessary identification information while keeping other private data secure? In the physical world, people can provide specific details when required (like showing a driver's license to the police) while keeping unrelated personal information private, such as their votes on police reform, donations to Planned Parenthood, or their work with the United Nations Relief and Works Agency for Palestine Refugees. But, how can we ensure we reveal only the required data needed in the digital world and no more?

State-run internet-based systems designed to solve these problems are problematic. In the last ten years, hundreds of government databases have exposed the data of billions of people.[18] Yet these centralized systems place immense power in the hands of bureaucrats, who can restrict access, erase credentials, or leverage financial institutions, telecom providers, and tech firms for mass surveillance.[19] In 2025, concerns about the US Department of Government Efficiency's access to sensitive IRS and Treasury Department databases have only heightened fears that such power can be abused, compromising personal privacy and democratic accountability.

Far from being citizen-centric, government identity systems often exclude society's most vulnerable—the poor, the rural, the homeless, the incarcerated, and the overworked—deepening existing inequalities.[20] A government-controlled identification infrastructure is not only impractical but also philosophically flawed. Why should any government have the ultimate authority to define our identities while maintaining the ability to expose, monitor, or manipulate them at will?

Instead, we need a self-sovereign identification system where each of us owns and controls our personal identifiers so that we can steward an accurate, safe, and privacy-protected identification for ourselves and others. A truly sovereign digital identity must be universally recognized: valid in both physical and digital spaces, across borders, and free from the control of any single entity. No corporation, government, or centralized authority should have the power to grant, revoke, or manipulate it. Such a system must be resilient, designed to outlast shifting political regimes, corporate leadership changes, and even its users, ensuring both continuity and the fundamental right to be forgotten.[21]

To establish this vision, identity must be decentralized and governed by the individuals it serves rather than external institutions. A self-sustaining identity framework would function as a shared digital commons, where users have explicit rights to safeguard their personal data, control how and where it is used, and determine its economic value. This commons-based model would empower individuals to define the principles governing their identities, fostering a system that prioritizes user autonomy, security, and collective stewardship over corporate or governmental interests.[22]

The next generation of the internet will be built on decentralized systems like blockchain. This iteration of the web is often referred to as Web3, and this technology is critical. Individuals must be able to store their data in the digital equivalent of a black box. Consider Civic, a blockchain-based platform that people can use to manage and verify digital identities in a secure and decentralized manner.[23] Founded in 2015, the company treats identity as a fundamental human right that should be universally accessible. Civic addresses the increasing risks of identity theft and fraud by giving users control over their personal information and tools for selectively sharing it with others.

One such tool is the Civic Pass, which issues a nontransferable "soulbound token" (SBT) that no one can move from its creator's wallet.[24] Civic Pass runs on more than ten blockchains and serves as a verifiable proof of a user's identity or of specific attributes, following verification processes like CAPTCHA, liveness checks, uniqueness checks, and ID document validation.[25] This approach not only enhances security and privacy but also facilitates trust and transparency in digital interactions. Developers can integrate Civic Pass into a variety of applications, such as airdrops, decentralized autonomous organizations (DAOs), decentralized finance (DeFi) platforms, real-world asset management, and smart contract development to enforce access control, prevent fraud, and ensure compliance.[26] It's proof that this is really happening.

## Stage II: From Identification to Digital Identity

As we have explained, identity is more than simple identification, like knowing who a person is and whether they are who they say they are. Identity is a complex social, cultural, and psychological construct, of which identifiers are only a small part. Support has been growing for the idea of a complete digital identity that consolidates all our data into a digital version of ourselves, a version that we own and manage in a digital wallet.

Leaders working on digital identities solutions like Phil Windley, for example, have argued that our data should be owned by us. "The more we rely on traditional identity systems, with their tendency to centralization, the more we risk ceding control over our digital lives, reducing our autonomy, and increasing our dependence on entities whose goals may not align with our own," writes Windley. Like us, Windley is a proponent of self-sovereign identity solutions, "where individuals maintain control over their own identity, reducing the risks associated with centralized access control and enhancing personal freedom in the digital realm."[27]

Take healthcare, for example. Our medical records are a core part of our identity. Our bodies generate biological data. We produce a heart rate and body temperature, not big companies, hospitals, or governments.

When clinicians measure us or run tests, they're capturing data from our bodies. With the rise of wearables and the Internet of Things (IoT)—a growing network of interconnected devices that gather and share data—we can now collect even more of this data ourselves, tracking everything from insulin levels and blood pressure to body temperature and daily steps.

Today, some hospitals have taken a step toward making medical records available to patients. Toronto General, which is part of the University Health Network (UHN) in Toronto, is ranked as the third-best hospital in the world.[28] If you go to its emergency ward with a broken arm, you can view your X-rays and radiology report before leaving the hospital.

Now, imagine viewing and owning that medical record as part of your self-sovereign identity. You could access it through a digital wallet and then share it with another hospital for a second opinion. You could choose to anonymize it, sell it, or donate it to scientific research. The point is that you set rules for its use, like allowing emergency medical practitioners to access it if you're found unconscious.

You could apply rules to any data so that, say, your driver's license reveals only that you've passed your driving test or that you're at least twenty-one years old *without* disclosing details irrelevant to the decision at hand. You could release only the specific information required for each situation while hoovering up and protecting the new data you generate as you navigate the digital world.

Blockchain technologies make this possible. Joe Lubin, CEO of ConsenSys, referred to this concept as a "persistent digital ID and persona" on a blockchain. "I show a different aspect of myself to my college friends compared to when I am speaking at the Chicago Fed," he said. "In the online digital economy, I will represent my various aspects and interact in that world from the platform of different personas." Lubin expects to have a "canonical persona," the version of him who pays taxes, obtains loans, and gets insurance. "I will have perhaps a business persona and a family persona to separate the concerns that I choose to link to my canonical persona. I may have a gamer persona that I don't want linked to my business persona. I might even have a dark web persona that is never linkable to the others."[29]

Your digital wallet may include information such as a government-issued ID, social security number, birth certificate, medical information, utility service accounts, financial accounts, diplomas, practice licenses, and various other credentials. It might also include sensitive data, like sexual preference or medical conditions, that you may not want to share openly but could choose to monetize in research studies or surveys. You could license these data for specific purposes, with clear rules for who can access it, when, and for how long.

## Stage III: Identic AI—the Virtual You Becomes Smart

Until now, the concept self-sovereign digital identity has focused on data, and the simple algorithms governing its security and use. But with identic AI, we stand on the verge of the most profound shift in identity yet. Our digital personae, once static repositories of information, are evolving into intelligent agents that can assist us and even act on our behalf. These agents can be trained to reflect our values, carry out tasks with autonomy, and serve as active partners in improving how we live and work.

Just as digital feudalism has reshaped our understanding of data ownership and control, AI companions and autonomous agents are transforming how we present ourselves to the world—and how we perceive our roles and potential within it. These AI agents will redefine daily life and revolutionize how businesses, governments, and other public institutions operate.

Beyond augmenting our knowledge and intelligence, AI will also extend and preserve our memories, even when they are lost catastrophically. "Memories are the architects of our identity," says technologist Pau Aleikum Garcia.[30] And yet they are fragile. Photos and records can disappear in political turmoil or natural disasters, while diseases like

Alzheimer's can strip individuals of their past. Garcia proposes a novel solution: synthetic memories—AI-generated, dreamlike visualizations of long-lost moments. These reconstructions could help reconnect families, restore a sense of self, or even enhance cognitive abilities in profound new ways.

The idea that intelligent digital agents can extend our consciousness and capabilities once belonged in the realm of science fiction. But with AI and other Web3, it is rapidly becoming reality. As AI systems become smarter and more sophisticated, we must confront profound questions about identity: Who are we, and what does it mean to be a "person" in a world where humans and their agents coexist? Could our digital selves evolve independently? Could they serve as role models, reflecting our potential and guiding us toward who we could become?

Sinead Bovell is CEO of the AI strategic insight and education firm Waye and an influential thought leader with hundreds of thousands of followers on social media. She calls the personal agents an "iphone moment" in the evolution of intelligent systems. "It's hard for people to understand how important a technology can be until it becomes personal. With the iphone, everyone had technology in their pocket and with them all day," she said. "The same is true for identic AI, as this technology is becoming very personal, not just something you use but an extension of you."[31]

Identic AI will inevitably cause us to reevaluate the criteria for personhood and reconsider the boundaries of human identity. If we consider consciousness and self-awareness as essential attributes of personhood, then we must ask: Could we someday classify advanced AI systems—those capable of learning, adapting, and even displaying emotional responses—as "persons"? This is not just a philosophical question; it raises pressing legal questions that challenge our frameworks for agency, intellectual property, and tort law. It also complicates our definitions of capacity and intent—concepts traditionally applied to humans but now tested by artificial entities.

What happens if our digital selves gain the ability to act on our behalf in civic matters, such as voting? Who decides when, or if, our AI representatives should be granted or stripped of such rights? No doubt many such questions are coming to your mind, too.

What happens when our identic agent becomes sentient, an event some predict could occur as early as the 2030s?[32] If our agent is capable of consciousness, could we upload our own one day?[33] If that were possible, it would mean a form of digital immortality, with our minds persisting beyond death. Could the ultimate breakthrough of identic AI be life after death? And if so, could we manage or even own our digital selves after death? Should we designate an heir for our virtual selves, much as we do with intellectual property rights?

## FROM IDENTIC AI TO TRANSHUMANISM

As we grapple with the evolving nature of digital personhood, we are forced to confront an even more profound question: Where lies the boundary between human and machine? If identic AI can extend our cognition, preserve our memories, and even simulate aspects of our consciousness, is it merely a tool, or the first step toward something more? The possibility of digital immortality raises unsettling yet fascinating dilemmas about identity, autonomy, and what it means to be human.

These questions are not new. For decades, the transhumanism movement has explored how technology can expand human potential, blurring the lines between biology and artificial intelligence. Advocates of transhumanism champion what they call "morphological freedom," the right to enhance our bodies and minds through technology.[34] While today's political climate highlights the ongoing struggle for bodily autonomy in areas like reproductive rights and gender identity, transhumanists envision a future where genetic engineering, AI, and cybernetic augmentation allow us to transcend our biological limits, extending lifespan, cognition, and human capability in unprecedented ways.[35]

Can identic AI augment human intelligence and creativity to the point of ushering in a post-human era—one where those with sufficient means can fundamentally alter their experience of being human? Some transhumanists argue that people could so profoundly reshape their physical and cognitive capabilities that individuals may no longer identify as human in the traditional sense. If that happens, would we still recognize them as human at all?

Such transformations pose profound risks. The history of our own genus offers a cautionary tale. Neanderthals, despite interbreeding with Homo sapiens 55,000 years ago, ultimately disappeared in the competition for survival.[36] Could a similar dynamic emerge between unenhanced humans and those who embrace radical augmentation? This brings us back to the Ship of Theseus: If we could transfer aspects of a person, like memories, thought patterns, or even consciousness, to a machine, at what point would we cease to be ourselves? How much could we replace with AI before losing the essence of personhood? And if that transfer were complete, would the machine inherit personhood in the process?

As we inch toward what Ray Kurzweil called the "singularity," we must rethink what defines humanity, what distinguishes our identities, and whether those identities might one day transcend our biological forms. This unprecedented technological ability to create digital versions of ourselves carries immense potential, but it also forces us to confront complex questions about trust, transparency, authenticity, and the boundaries between our real and virtual selves.

In the pre-AI world, we each plan and navigate our lives, "designing our future," so to speak. But what happens when we must design for two: for ourselves and our AI counterpart? And why stop at just one AI self? What if we could run multiple virtual versions of ourselves, each exploring different scenarios and possibilities simultaneously? Could we embrace such a "multimodal" existence without losing our sense of self . . . or our sanity?

As we delegate tasks and decisions to AI, we must first define the values and the principles that guide our lives. What information should we retain in our memory versus offloading to our AI agent? Which activities, tasks, and challenges should remain ours, and which should we entrust to a digital self? Where do we draw the line on the problems our agent can tackle?

Beyond functionality, we must also consider the deeper implications of outsourcing aspects of our identity. How will our agent present us to the rest of the world? Will relying on AI erode essential skills, and if so, does that matter? If our AI companion makes decisions or solves problems for us, might we sacrifice our critical thinking and creativity?

For your digital self to contribute meaningfully to your life, it must have access to comprehensive and intimate details about you, such as your health data, purchasing habits, social interactions, and behavioral patterns. In other words, it would need the kind of deep personal insights that big tech firms already collect. (Remember how Target predicted a teenager's pregnancy before her own father knew?[37])

However, as we explore in chapter 3, large corporations currently control the AI data and foundation models that power these systems. Without a fundamentally new approach to AI, we risk enabling unchecked tracking, profiling, and exploitation. As with social media, we may once again trade our privacy for convenience, unaware of the long-term consequences.

Beyond privacy concerns, our AI companions will have instant access to much of the world's knowledge, essentially becoming "know-it-alls" in the most literal sense. Will we really want an insufferable virtual version of ourselves? More important, identic AI—both ours and everybody else's—derives its knowledge from the world's datasets, which often contain inherent biases. If left unchecked, could our digital copilots perpetuate discrimination in hiring, management, or decision-making? The risks are not just personal but societal, shaping the very systems we rely on.

## THE DIGITAL SELF IN SOCIETY

Our identity is not simply a personal memory or self-reflection; it is our positioning in society. We each have a collective identity linked to memberships in various groups. We're proud Texans, trade unionists, Chicago Cubs fans, crypto developers, TikTok influencers, and members of book clubs and other congregations. How will we design our digital agents to engage more deeply in these communities?

When we curate our social media identities, we often portray idealized versions of ourselves. How authentic are we today? Are we conforming to social expectations? Our Facebook, Instagram, WeChat, Snapchat, Tik-Tok, and LinkedIn profiles and posts corroborate this ideal version. We're having fun, attending the best concerts and parties, eating delicious food, and hanging out with desirable people. Are we being true to ourselves in the digital realm, or are we merely projecting a carefully constructed version of our identity?

Now imagine when the self you're projecting is not simply a two-dimensional photo or video but an intelligent and potentially sentient digital entity representing the various modes of your life. How can we be sure you are who your agent says you are? Or will you even want it to? Will you lose your authentic self while your digital self is rocking in the free world? Erving Goffman must be rolling in his grave.

Locke argued that personal identity was grounded in consciousness, particularly memory; our ability to remember and reflect on past experiences defined our identities. What if we had a perfect memory of everything? Would a supercharged "continuous sense of self-awareness" unlock a profound new kind of human identity? Or would we become more insufferable? Your virtual know-it-all arguing with mine or agreeing on everything, and so we'll have nothing to discuss except how much we've forgotten? What makes human beings interesting is their messiness, imperfections, screw ups, and bounce backs. Come to think of it, that's what makes life pretty interesting, too.

Remember the questions posed by the Greek philosophers. What defines an identity? Is an object the sum of its parts, or is there an underlying "essence" that persists despite changes? And is there a threshold of digital change or enhancement beyond which our authentic identity is lost?

Now, buckle up because identic AI is set to challenge these very ideas. It blurs the boundaries between our physical and virtual identities, reshaping our understanding of identity in profound ways. As we navigate this new reality, we must critically examine how we assert control over our representation in an increasingly AI-driven world. The evolution of identity in the digital age will demand careful curation of our digital selves, ensuring they align with our authentic selves while preserving our agency to shape who we are in both realms.

Ultimately, true agency comes from a commitment to human-centered values, prioritizing relationships, autonomy, well-being, and meaning over efficiency. If we are to harness the potential of identic AI responsibly, we must design our digital identities to reflect and uphold our core human principles. The future of identity is not just an individual choice but a collective one—one that determines whether we realize the promise or succumb to the perils of identic AI.

Tough and complex issues, for sure. Let's start with the potential.

## Chapter 3

# THE TECHNOLOGY OF IDENTIC AI

**W**ithin a remarkably brief span of time, AI models have become amazingly capable. They have enhanced reasoning abilities and can perform a wide range of useful functions. They can interact and respond in real time. And, as adoption surges, AI will transform supply chains, upend financial systems, accelerate scientific research, and shape industries. The predictions about the coming impact sound hyperbolic, but they are not unthinkable. Says historian, philosopher, and author Yuval Noah Harari: "For the first time in history, we have something that can shape culture, write poetry, and influence politics without human input. This is more powerful than any technology we've ever created."[1] The next decade promises rapid progress, new challenges, and opportunities that we are only beginning to comprehend.

You're not alone if you find this shift unsettling or overwhelming. Technology is no longer evolving at the speed of the internet; it's moving at the speed of AI. No wonder so many of us feel disoriented, struggling to keep pace with the stealth force reshaping our world. Bob Dylan nailed the feeling: "There is something going on here, but you don't know what it is, do you, Mister Jones."[2] So let's pause for a moment, take a deep breath, and consider the transformation unfolding before our eyes.

## THE RISE OF IDENTIC AI

By now, most readers of this book will have experimented with AI apps like ChatGPT, Siri, and Copilot. If you ask ChatGPT to summarize a document or write a report, its underlying large language model (LLM) can process your request and produce coherent and context-appropriate text. It can solve equations, propose software code, and even automate workflows.

But the field of AI has made a critical leap: AI can now have agency, meaning that it can anticipate its users' needs, plan actions, and optimize outcomes in real time. AI functions with a level of autonomy in which it can improve processes and make decisions without requiring constant human oversight.

Agency defines the next generation of AI. Among these "agentic AI" systems, identic AI stands apart. These aren't just general-purpose agents like those in call centers or on supply chains, or simple office assistants that perform limited and repetitive tasks. They are deeply personalized digital counterparts, capable of extending their humans' capabilities. Identic agents not only assist us; they can manage our daily lives, enhance our intelligence, and promote our well-being, evolving alongside us to become indispensable partners in our work, our healthcare, and our personal growth. In short, identic agents turn our digital identities into powerful extensions of ourselves.

To distinguish between LLMs, agentic AI, and identic AI, let's consider two examples: an office assistant managing a calendar and a customer service operation undergoing automation.

In an office setting, an LLM-based assistant awaits explicit commands.

You ask it to schedule a meeting, and it does—nothing more. It doesn't anticipate conflicts or proactively manage your workload. General-purpose AI agents go further. They can analyze workloads, predict scheduling conflicts, and prioritize tasks to optimize workflow efficiency. Identic AI is even more advanced: It understands your habits, work style, and stress patterns, and it adjusts your calendar accordingly, perhaps rescheduling a meeting when it detects your weariness, reminding you to take a break, suggesting delegates for you, analyzing strategies to adopt in the meeting, or drafting your meeting talking points.

Now, consider a customer service operation that seeks to modernize. An LLM-powered chatbot can respond to customer inquiries efficiently, but it lacks strategic awareness. It cannot optimize a company's entire customer service system without human oversight or modify workflows based on evolving business needs.

By contrast, an agentic AI system operates with a greater autonomy. It predicts demand, reallocates resources, and adjusts workflows dynamically. Unlike an LLM, it doesn't need constant human input to improve operations; it actively optimizes them in real time.

Identic AI takes this autonomy a step further by integrating personalization and emotional intelligence. Let's say your employer just eliminated your customer service role. Instead of leaving you to navigate an uncertain future alone, your identic agent steps in as a career strategist, skills coach, and transition guide. It analyzes your strengths, identifies new career opportunities, and curates a personalized upskilling plan. Recognizing your strong analytical and communication skills, it might suggest a transition into customer experience management, recommend tailored courses, set up networking opportunities, and even conduct mock interviews to prepare you for your next role.

This level of autonomy opens new avenues for productivity, personalization, and innovation. Beyond the workplace, the opportunities are, well, limitless. Just think about what happens when our agents are smarter than us. Or when we can replicate agents to take on countless life challenges for each of us as Peter Diamandis wants to do with Peter Bot. He argues that his capability and powers will grow exponentially as he assigns his digital self to take on countless challenges. He muses about what people one hundred years ago would think of us with our digital superpowers in tow. "To past generations we would be seen as having magic. We would be gods."[3]

Of course, the proposition of becoming godlike raises important questions about control, ethics, and accountability—topics we'll explore later.

## HOW WILL YOU INTERACT WITH YOUR DIGITAL SELF?

Whether at work or in everyday life, how we interact with identic AI will be almost as life changing as the technology itself. Today, most of us rely on familiar interfaces like smartphones, desktop computers, and voice-activated devices to access AI-powered tools like chatbots and virtual assistants. But as identic AI evolves into more proactive, autonomous systems, these interfaces will need to adapt, becoming more immersive, intuitive, and seamlessly integrated into our daily lives. In the future, our interactions with AI companions and copilots will be as natural as speaking to a friend, collaborating with a colleague, or sharing ideas in a brainstorming session.

### From Text to Multimodal Communication

Currently, most AI interactions are text based or voice driven, constrained by the devices we commonly use. People type questions into a chat window or issue commands to a voice assistant like Alexa or Siri. While functional, these interfaces are limited in facilitating deeper, context-rich collaboration. Identic AI systems, however, can do more than react; they are designed to engage, adapt, and intervene.

To achieve this, they will rely on multimodal context-aware communication, combining text, voice, visual data, and even gestures to create richer interactions.

Instead of responding to typed prompts or verbal commands at work, your identic agent will integrate seamlessly into your workflow. Using a combination of real-time video feeds, voice recognition, and augmented reality overlays, the AI will highlight key points in a presentation, pull up relevant datasets during a meeting, or suggest edits to a document as you write. For example, a software engineer in a virtual meeting might use their AI copilot to display a dynamic, 3D visualization of a new user interface concept, adapting in real time to questions and engagement levels from their peers.

## The Rise of Holographic and Augmented Reality Interfaces

One of the most significant leaps in human-AI interaction will come from the adoption of holographic and augmented reality (AR) interfaces. These technologies allow identic AI systems to step out of screens and into our physical environments, creating a more immersive and collaborative experience. Soon, your digital self will appear as a holographic presence in company meetings, sitting alongside team members, offering real-time insights, and even interacting with other AIs representing colleagues or departments. This holographic copilot could annotate a shared virtual whiteboard, display 3D models of a product in development, simulate market scenarios on demand, and offer a critical insight.

In everyday life, AI companions might appear as holographic guides embedded into AR glasses. Picture yourself walking into a bustling airport. Your AI companion projects navigation arrows, highlights your gate number, and suggests nearby dining options, all in your field of vision. Such interfaces will make identic AI feel less like a distant service provider and more like an ever-present partner that syncs up with the rhythms of your daily life.

## The Metaverse: AI as a Native Inhabitant

As the metaverse matures, it will become a significant arena for interacting with AI agents. For example, agents could act as native inhabitants in virtual environments to enhance human experiences within these spaces. Picture digital proxies—AI-powered avatars—for individuals in virtual meetings, negotiating contracts, collaborating on projects, or even brainstorming creative solutions in real time. These AI agents, acting on behalf of their human counterparts, could analyze data, translate languages, and offer strategic insights, transforming how we socialize and conduct business within the metaverse.

In education, students will engage with AI companions in immersive virtual classrooms, where the AI not only acts as a tutor and research assistant but also curates personalized learning experiences. For example, Mary Lacity, the David D. Glass Chair and distinguished professor at the University of Arkansas, describes how the university has shifted almost all of its science curriculum to the metaverse, where students wear virtual reality headsets and interact with AI teaching assistants embodied as virtual avatars. "Our students are so digitally native in this three-dimensional world that I find myself learning a lot from them," said Lacity. "They come up with ideas, create learning experiences, and develop projects that absolutely blow my mind."[4]

## Voice, Gestures, and Neural Interfaces

Interfaces will evolve beyond augmented reality and the metaverse. As technology advances, people will interact through gestures and neural interfaces. With gesture recognition, already used in gaming and smart home systems, users could interact with AI by pointing, waving, or gesturing otherwise. For example, with a simple wave, a factory worker might prompt an AI-powered robotic arm to adjust its position or pause operations.

Neural interfaces, or devices that read and interpret brain signals, offer an even more transformative possibility. Although still in their infancy, these systems could allow humans to communicate directly with identic AI through thought alone. Imagine a healthcare professional who can summon patient data or diagnostic recommendations with a mental cue, or a writer who can "think" their ideas onto a page, guided by an AI that refines their prose in real time. While such technologies are in the exploratory phases, they hold the promise of making AI an almost seamless extension of human cognition.

## The Path to Invisible Interfaces and Digital-Physical Integration

Ultimately, the goal of identic AI is not to create better tools but to mesh with the fabric of our lives so that we can focus on what matters most. As interfaces evolve, they will likely become less visible, shifting from physical devices to ambient systems that respond to context and intent. AI companions could run through smart home ecosystems, wearable devices, or even embedded sensors, offering assistance without explicit commands. For example, your agent might read your biometric sensor and remind you to hydrate or offer a mindfulness exercise, without your asking. As these technologies mature, the boundaries between our human selves and our identic AI selves may disappear.

Some of these agents may even take physical form, and over time most will. The rise of robotics is moving intelligent systems from screens into embodied reality. We're beginning to see AI agents manifest as autonomous service robots capable of engaging in social interactions, assisting with care, delivering food, managing inventory, or guiding people through unfamiliar spaces. In the near future, it's likely that your identic AI will have a physical avatar: a robotic embodiment capable of standing in for you at a meeting, greeting guests at a venue, accompanying you for a walk, or performing errands on your behalf. These robotic agents won't just be tools; they'll be emissaries of your digital self, blending cognition, personality, and presence into a seamless extension of who you are.

Sound far-fetched? Surely personal robots would be too costly for all but the super rich. Besides, today's robots don't do much, and they often appear foreign and even unsettling to many of us. But so did previous seismic technological innovations, like cars and computers, in the early stages of development. Consider the trajectory of the automobile. When cars first appeared, they were expensive, unreliable, and viewed as noisy playthings for the wealthy. Roads weren't built for them, laws weren't prepared, and few people could see their value. Gradually, though, infrastructure improved, costs dropped, and utility rose. Today, the car is so integral to daily life that the family vehicle often has its own room in the house: the garage.

Robots are at a similar inflection point. At present, they're costly, limited in function, and the subject of ongoing debates about safety and ethics. But as identic AI advances, and as we overcome design, regulatory, and trust barriers, their utility will soar. With more data, they'll grow smarter and eventually merge with their purely digital counterparts. Economies of scale will drive down costs. And just as society adapted to accommodate cars, we'll adapt to welcome physical identic agents. They too may earn a place in our homes and buildings—not in a traditional garage, but perhaps in a dedicated corner, chair, or closet within our living spaces.

The rise of personal robotic agents brings profound challenges that we must confront directly if this remarkable innovation is to evolve safely. Above all, we must safeguard human agency in a world where our digital selves take physical form. It's concerning enough when a virtual assistant misunderstands or defies our intent; the risks are magnified when that agent walks in titanium and wields the strength of gods.

## HOW DID WE GET HERE AND WHERE ARE WE GOING?

In 1997, IBM's Deep Blue defeated chess grandmaster Garry Kasparov. "I asked Garry what happened," recalls Manoj Saxena, reflecting on his time as general manager of IBM Watson. "And he said, 'Typically a chess grandmaster can see twenty or twenty-two moves deep. And that makes me a superhuman when it comes to chess. But there were times when Deep Blue was looking ninety-five moves deep. We didn't even know those moves existed.'"[5]

Then, in March 2016, with over two hundred million people watching online, the legendary Lee Sedol played a game of Go with an unorthodox opponent: an artificial intelligence developed by DeepMind, Google's AI subsidiary. The decade's greatest player, renowned for his deep strategic thinking and adaptability, Sedol had won eighteen world titles. His opponent AlphaGo, the product of advanced neural networks and reinforcement learning, trained on millions of games.

Few believed that a machine, no matter how skilled and sophisticated, could best a human master at Go, an ancient board game known for its complexity and vast number of possible game configurations. Opponents play Go on a nineteen-by-nineteen grid, almost six times greater than the eight-by-eight grid of chess. As a result, there are approximately $2.1 \times 10^{170}$ possible legal Go positions compared to about $10^{40}$ positions in chess. Even if you had every exaflop supercomputer humanity could ever build, you'd need on the order of $6.7 \times 10^{144}$ years to step through every legal position of Go. The Universe is $14 \times 10^{10}$ years old, so it would take all these supercomputers $4.8 \times 10^{134}$ universe lifetimes to "solve" Go. In comparison, the age of the Universe is so small that even if you had a trillion universes existing sequentially, one after the other, you would still need another approximately $4.8 \times 10^{134}$ to finish the computation.

So most experts assumed that even the most powerful AI systems would need far too much computing time to devise winning strategies in a game environment. In practice, computer programs could only play, at best, a mediocre amateur game. Few were prepared for what happened next.

Early in the game with Sedol, AlphaGo played a move—Move 37—that defied conventional Go wisdom and visibly stunned Lee Sedol, the game commentators, and the global audience of Go experts. AlphaGo placed the move on the fifth line, an unusually high position. Traditionally, top human players would have considered such a highly unorthodox move too bold and potentially reckless.

Move 37's significance lies in its departure from established Go strategies. Human players rely on centuries of accumulated knowledge and pattern recognition, often leading to conventional moves that adhere to tried-and-true tactics. AlphaGo's move showed that AI could transcend human constraints and explore strategies that human players had not previously considered.

Move 37 forced Sedol to rethink his strategy and adapt to AlphaGo's unconventional approach. Despite his best efforts, Sedol lost the game, citing this move as a crucial factor in his defeat.

Move 37 demonstrated AlphaGo's ability to challenge even the most skilled human players and showcased the potential for AI to push the boundaries of human knowledge. Sedol later described the experience as profoundly humbling, saying it revealed a new dimension of Go—and of intelligence itself.[6] Most experts saw this breakthrough as a "decade ahead of its time."[7]

The implications of Move 37 extend far beyond the realm of Go. After processing a huge amount of data and learning, AI came up with novel solutions not immediately apparent to humans. Could AI revolutionize other fields that require deep expertise, strategic thinking, and problem-solving?

At the same time, people speculated on AI's potential in creative domains. If AI could act so creatively in a strategic game, perhaps machines could contribute to creative processes in art, music, and literature. Ultimately, Move 37 secured AlphaGo's victory and everyone's interest in the impact of intelligent machines on our world.

## Large Language Models: A Big Leap Toward Identic AI

Whereas DeepMind designed AlphaGo to master one complex game, the subsequent development of LLMs such as OpenAI's GPT-4 (generative pretrained transformer) have revolutionized AI's ability to understand, generate, and interact with human language across countless domains. How quickly this new generation of AI models has evolved!

OpenAI's GPT-2, released in 2019, demonstrated how a model trained on massive amounts of text data could perform surprisingly well on various natural language processing (NLP) tasks, such as text generation, summarization, and even answering questions. However, in 2020, the release of GPT-3 truly showcased the power of LLMs. With 175 billion parameters, GPT-3 could generate humanlike text, from writing essays to coding simple programs.[8]

GPT-3's ability to generalize across tasks with little instruction showed that AI could perform specific tasks and exhibit a degree of flexibility that approached humanlike learning. For the first time, AI could assist in real-time decision-making, conduct complex research, and interact more meaningfully with users across industries like education, customer service, and creative content generation. In many cases, GPT-3's text-based outputs were so good that they were almost indistinguishable from human creations.

While GPT-3 broke ground with 175 billion parameters, a new generation of models—like OpenAI's GPT-4o, Anthropic's Claude Sonnet 3.5, Google's Gemini 1.5, and Llama 3.2 from Meta—have leapt forward with enhancements in scale and sophistication. For example, GPT-4 can now handle multimodal inputs, meaning that it can process text and images. With multimodes, AI applications can do more, such as interpreting visual data and responding appropriately in context across domains like design, diagnostics, and customer service.

In fields like medicine, multimodal capabilities allow AIs to read patient charts while interpreting medical images. In education, students can interact with AI tutors that are capable of visually and verbally explaining concepts. Meanwhile, in agriculture, an AI copilot could identify early signs of crop disease by processing visual and spectral data from drones and cross-referencing it with soil moisture levels. Farmers would have more time for mitigation.

Another advance in GPT-4 was its enhanced accuracy and reasoning abilities. Even though GPT-3 could generate humanlike text, it wasn't always accurate or coherent in complex discussions. GPT-4 addressed these shortcomings with better context retention and more nuanced responses, making it more reliable for tasks requiring in-depth analysis or decision-making. For instance, GPT-4 performs better in areas like legal reasoning, scientific analysis, and programming, where consistency and precision are critical.

Moreover, OpenAI has trained GPT-4 on even more diverse datasets, so that it can excel in multiple languages and adapt to more professional and creative tasks. These advances have made it an even more versatile tool for industries ranging from healthcare and education to content creation and software development. In short, GPT-4 not only builds on the groundbreaking NLP abilities of GPT-3 but extends the frontier, taking us closer to truly intelligent AI systems that can act as powerful copilots in professional and personal settings.

Ethan Mollick, author of *Co-Intelligence: Living and Working with AI*, observed, that "if GPT-4 level performance was the maximum an AI could achieve, that would likely be enough for us to have five to ten years of continued change as we got used to their capabilities."[9] Of course, we see no imminent slowdown in AI development on the horizon. On the contrary, yet another new generation of models was emerging as we were putting finishing touches on this manuscript.

In 2025, models such as OpenAI o1 and o3-mini were a significant leap forward in AI's multistep "deliberative reasoning" before generating responses. Unlike earlier models that produce answers almost instantaneously, these models allocate additional processing time to analyze complex queries, weigh potential solutions, develop a chain of thought, and refine their output. With this extra "reasoning," these models can tackle challenges that require deeper logical analysis and multistep problem-solving.

According to scientific research, o3-mini can address PhD-level problems across such disciplines as advanced mathematics, physics, and biomedicine.[10] Researchers have used the model for intricate calculations, hypothesis testing, and interpreting nuanced academic material. In this capacity, o3-mini functions not just as an automation tool but as a collaborative assistant.

## The DeepSeek R1 Disruption

In early 2025, the small Chinese start-up DeepSeek quietly released a new AI model. Within days, whispers turned into headlines, and headlines turned into a reckoning.

The model, called R1, was more than good. It was groundbreaking. It matched or outperformed the best frontier AI models of industry giants with far deeper pockets. That alone impressed us. What shook the AI community was how R1 achieved its high level of performance.

The cost to train R1 was a fraction of what its competitors had spent. Even more astonishing, its run-time inference costs—the expense of using the model—were dramatically lower than everyone else's. Raw computational power no longer determined success. DeepSeek had rewritten the rules.

At the heart of R1's efficiency was its model's radically different approach to processing information. Instead of functioning like a massive monolithic system, where every part of the model had to participate in every task, R1 worked more like a network of specialists.

It used a technique called *mixture of experts* (MoE). Instead of asking a single AI model to "know everything," requiring immense computational resources, DeepSeek designed R1 more like a team of domain experts, where each specialized in a different area of knowledge. When a user posed a question, the model didn't engage all 671 billion of its parameters. Instead, it activated only the subset of experts most relevant to the task, sometimes using as little as thirty billion parameters at a time.

The result? A model that was faster and cheaper to run and that scaled more intelligently.

DeepSeek didn't stop there. It trained R1 largely on synthetic data rather than expensive, time-consuming, and inherently limited human-created datasets. DeepSeek wasn't hoovering up data over the internet or other channels; it was generating, testing, and refining its own training data. Through reinforcement learning, R1 essentially trained itself on the output of other models to improve rapidly. It accelerated what once took months or even years to refine.

In its pursuit of greater efficiency, DeepSeek also used a process called *distillation*. Rather than foist every use case on the R1 model, DeepSeek developed its larger models to train smaller models, compressing its knowledge into smaller, highly efficient versions. These "student" models, some with as little as 1.5 billion parameters, retained extraordinary reasoning ability while slashing computational costs.

The implications were immediate. If DeepSeek could build a smaller, faster, and dramatically cheaper AI that performed as well as the most advanced frontier models, then the future of AI wouldn't belong to a few tech giants with unlimited resources. The barriers were collapsing.

Almost overnight, the industry began a new race, one based not on scarcity and cost but on abundance, commoditization, frugality, and ubiquity. In other words, the AI arms race was no longer just about training ever-smarter models; it was about diffusing these models everywhere.

We are entering a new period of human culture. In this period, our relationships with humankind and machines, and our assumptions about technology, will change dramatically.

## Artificial General Intelligence: Personal Superintelligence

Each step in the evolution of machine learning technologies and agents has brought us closer to realizing the dream of truly autonomous systems. Soon, AI will understand, learn, and apply knowledge at a level comparable to or exceeding human intelligence—often referred to as *artificial general intelligence* (AGI)—across a wide range of tasks. When this occurs, and if we do this right, our identic agents will truly extend our cognition and capabilities, seamlessly integrating into our lives.

Today, AI has surpassed human capabilities in specific domains, such as game-playing, pattern recognition, and data analysis. However, the journey to AGI represents a more profound shift, where machines operate at a level of cognitive ability akin to humans. On the next frontier, AI developers will create systems that will not only solve problems within defined tasks and goals but will also learn, adapt, and apply knowledge across entirely new domains. Applied to identic AI, your agent will reason, improve itself, intuit, and interact socially like you. This transition from AI to AGI is the holy grail of AI research and promises to reshape our world in ways we are only beginning to imagine.

In 2025, OpenAI's GPT-4 and similar models brought AI closer to achieving human-level proficiency in understanding language, solving problems, and generating creative outputs. OpenAI's o3-mini and Deep-Seek's R1 added advanced deep multistep chain-of-thought reasoning, which was a better bridge to human-level cognition. DeepSeek trained these models on enormous amounts of data so that they could make real-time adjustments and learn from interactions. Yet, Manoj Saxena and others believe these advances tap the surface only.

"The shift we are about to make is bigger than the internet," said Saxena, formerly of IBM. "It's bigger than previous era–defining inventions like fire, electricity, and refrigeration. This is a fundamental shift in amplifying our knowledge and our intelligence, much like what the steam engine did for our arms and legs. This is going to amplify our minds, and it is going to make us go beyond superhuman."[11]

Augmented superhuman intelligence will eventually push the frontiers of human knowledge and scientific discovery, including what we presently cannot conceive. "Using AI to explore the unknown unknowns is where the real exciting possibilities for humanity are," Saxena said, referencing the Rumsfeld Matrix.[12]

To get to AGI, however, we must do more than scale current AI models. We must develop machines that possess:

- *deep understanding* of the world, that is, the "knowledge and insight gained through in-depth analysis and study,"[13]
- *commonsense reasoning*, the ability to make deductions and reach conclusions using knowledge about everyday objects and their interactions,[14] and
- *transfer learning*, the ability to transfer and apply knowledge from one domain to another without explicit programming.

On these dimensions, current AI systems still fall short. To make progress, scientists will likely combine different AI techniques, including symbolic reasoning with neural networks and advances in neuroscience-inspired algorithms.

Achieving AGI is not just a matter of smarter algorithms; it also depends on access to big data and computational resources. Over the last decade, the growth of big data—massive, diverse datasets collected from countless sources—has fueled advances in AI. Companies like Open AI, Google, Microsoft, and IBM are leveraging big data to train AI systems so that they learn from patterns in text, images, and other sensory data. This data availability is critical to AGI's comprehensive understanding of the world.

The International Data Corporation (IDC) predicts the global AI infrastructure market will surpass $100 billion in spending by 2028.[15] Some estimates are much higher.[16] Meanwhile, traditional chip manufacturers like NVIDIA, AMD, and Intel are developing advanced *graphical processing units* (GPUs) optimized for AI workloads while Google, Apple, and Tesla are investing significantly into AI silicon.

While we will see extraordinary advances in AI algorithms and AI-optimized chips, we will also see major innovations in distributed computing and quantum technology. These technologies could very well provide the raw power that AGI needs to perform tasks far beyond the current limitations of AI, so that machines can process, learn, and reason at a scale and speed previously unimaginable. On December 9, 2024, Google launched Willow, a new quantum chip that achieved a random circuit sampling benchmark—the hardest benchmark that we can set for a quantum computer—in under five minutes. Such a computation would take one of today's fastest supercomputers ten septillion (that is, $10^{25}$) years to complete.[17] That vastly exceeds the age of the universe.

Despite these advances, vital challenges still stand between us and AGI. AI cannot generalize from limited data; that is a hallmark of human intelligence. Human beings can make sense of new situations and apply learned knowledge across multiple domains; AI systems still struggle to do so.

Creativity and emotional intelligence are also fundamental to identic agents. While AI can mimic creative processes, it cannot generate truly novel and contextually relevant ideas. Current AI lacks the intuitive leap that humans often make when they face a challenge. Similarly, AI lacks social and emotional intelligence, the abilities to understand, respond to, and engage with humans on social and emotional levels.

While the precise timeline for achieving AGI remains unclear, the focus of the AI community has shifted from "if" to "when." And the time horizon is rapidly closing. Some leading figures in the AI community have offered what many consider surprisingly near-term forecasts. Former Google CEO Eric Schmidt describes the "San Francisco consensus," where many leading thinkers in that city believe AGI will be here by 2030.[18] Ray Kurzweil, Google's Director of Engineering and a noted futurist, has long predicted that AGI will be achieved by 2029, with a full technological singularity, where AI surpasses human intelligence in all respects, arriving by 2045.[19] Others remain more skeptical. In a 2025 paper titled "The Illusion of Thinking," Apple researchers found large AI models show a "complete accuracy collapse" on complex problems and conclude that true AI reasoning remains far off.[20] A 2025 Pew Research Center survey reinforced this skepticism: Nearly half of AI experts (48 percent) said they believe it is unlikely AI will develop independent thought within the next twenty years.[21] But there is broad consensus that AGI is coming, and it's coming fast. As researchers continue to push the boundaries of machine learning, neuroscience, and cognitive science, the leap from narrow AI to general intelligence may be closer than we think.

# TOWARD THE SELF-SOVEREIGN DIGITAL YOU

As we journey toward AGI and the full realization of the digital self, another fundamental challenge emerges: ownership of the applications and infrastructure that enable the digital you. These identic agents rely on gigantic datasets, sophisticated algorithms, and extensive computational resources concentrated in the hands of a few powerful entities.

In chapter 2, we introduced a three-stage framework describing the evolution of the digital self, beginning with a self-sovereign system where individuals owned and controlled their digital credentials. We argued that consolidating medical, financial, and social data into a user-managed digital wallet laid the foundation for a full digital identity. With identic AI, these static repositories of information could evolve into intelligent agents—digital extensions of ourselves that assisted us and even acted on our behalf. We closed by arguing that we, and not Big Tech, should own our digital identities.

## Should Big Tech Own the Digital You?

At the close of the twentieth century, tech visionaries said the internet would unite peoples, cultures, and societies by democratizing information and freeing the flow of new ideas. It would be a great equalizer, a force for good. Today, technology behemoths such as Alphabet, Amazon, and Meta in the West, and Baidu, Alibaba, and Tencent in the East dominate the internet of the twenty-first century; and all have relationships with the men in power like Trump and Xi. These giants are more than corporations; they are a new species of business, the apex predators of all markets they pursue. Their scale and magnitude give them power that is both awesome and frightening. With their financial strength and access to limitless resources, they can expand rapidly, edge out competitors, and disrupt ecosystems that billions of consumers, workers, and entrepreneurs, as well as thousands of start-ups and other businesses, depend upon and once thrived in.[22] Increasingly, these colossi claim ownership to the primary asset class of the modern digital economy: data—theirs, yours, and ours.

This concentration of wealth and power raises critical questions about control, access, and equity. Emad Mostaque, founder and former CEO of Stability AI, asks, "Who builds the AI that teaches your kids, makes critical medical decisions, and delivers essential government services? It probably shouldn't be private market forces."[23] For sure, corporations ought to participate and negotiate for data, but full ownership of our digital selves should be off the table. Big technology companies such as Google and Meta built fortunes on advertising business models and behavioral influence. Both companies harvest and exploit user data like your browsing history and social network activity to target and tailor advertising. Their ability to profile users precisely makes their platforms very attractive to advertisers and politicians.

Among other goals, these companies will undoubtedly use AI to reinforce and expand their ad-driven ecosystems. AI chatbots and digital assistants will likely become conversational sales engines, subtly deploying psychological triggers to guide users toward sponsored recommendations and purchasing decisions. Ad-optimized model training could prioritize corporate-friendly narratives and fine-tune responses to favor advertisers' products. AI-generated dialogues could seamlessly weave in brand messaging—like subtly placing a "refreshing Coca-Cola" to trigger a craving. Or, imagine asking your identic AI to compare the self-driving features of three new cars you're considering. Unbeknownst to you, a car manufacturer has paid the AI platform to favor its vehicle. Your digital self is biased, and you can't even tell!

If we don't own our digital identities, Big Tech and its advertisers will gain an unprecedented, irrevocable, and largely invisible power to shape our preferences and our perceptions of reality. The next era of advertising won't sell just products; it could embed messages directly into our extended consciousness. This brings a frightening new meaning to the marketing practice of "product placement."

Meredith Whittaker, president of Signal and a leading voice on tech accountability, warns that these companies are not just building tools: They're constructing "surveillance business models" that depend on extracting behavioral data at massive scale. As she puts it, "The more intimate the AI, the more invasive the surveillance."[24] Identic agents, embedded into every aspect of life, could become the ultimate listening devices, normalizing continuous data capture under the guise of assistance. Without enforceable guardrails, the very agents designed to support us may instead become instruments of manipulation and control. For example, politicians, ideologues, election candidates, religious leaders, conspiracy theorists, enemies of democracy, or any other "advertiser" could plant their messages into our digital experiences, nudging us toward or even grooming us for their cause in ways we may never detect if the ad price was right and we fit the profile.

Moreover, if corporations or governments control the data and infrastructure, they effectively control the development and deployment of our digital selves at work, in education, healthcare, and countless other settings. They could stifle innovation, limit individual agency, and exacerbate societal inequalities. Conversely, decentralized and transparent approaches to data ownership and AI infrastructure could democratize access and serve the broader public good.

That's partly why we argue that you should own your data, identity, and digital agents to the greatest extent possible. More to the point, this digital you knows everything about you. As individuals and as a society, we need our data and our digital identities to be secure, private, under our control, and owned by each of us. In short, we need them to be "self-sovereign."

Indeed, the potential for AI to shape every aspect of society has a growing number of experts agreeing with us in calling for a shift away from centralized control to a more open, decentralized model. Harper Carroll, a widely followed influencer, Stanford AI graduate, and founding engineer of a company acquired by NVIDIA, explained why this shift is both necessary and achievable.

"Math is the universe's fundamental language, and these algorithms based on the human brain have universal applications across industries, problems, and social contexts," said Carroll.[25] Carroll says it doesn't make sense for such powerful extensions of cognition and expressions of universal principles to be centrally controlled or owned by a select few. "If a small group controls these algorithms, they effectively control everything since the math applies in any context where one can learn from data."[26]

Carroll suggests that if the architecture of AI is centralized, regulation could become necessary to avoid dangerous consequences, but she warns against strict regulatory frameworks that could stifle innovation. "Forcing the world's technological development through such narrow control mechanisms is fundamentally unsustainable and borders on absurd," said Carroll. "A better approach is open, decentralized, and democratic, which is achievable. With today's technology, individuals can train their own models independently."[27]

To date, however, the prevailing assumption has been that, like previous technological revolutions, massive companies will provide the technology to create and sustain self-sovereign digital selves in perpetuity. True, throughout modern history, corporations launched, developed, and ultimately captured the markets for energy, electricity, telephony, mobility, computing, the internet, and social media. True, the tech behemoths of today already offer you some type of AI to help you invest your money, pay your bills just in time, and find jobs, homes, or romantic partners.

These are likely governed by the Faustian bargain you've already struck with them. They will offer you these capabilities as part of an advertising, subscription, or sponsorship business model. What these companies will be unwilling and perhaps unable to give you is the "sovereign you." Given the huge costs of developing and continuously advancing AI models and systems, they will surely want to own the digital you.

Should governments intervene and act forcefully to break up these companies? Do we need new regulations to rein in the world's technology pioneers? Without question, people are increasingly dissatisfied with Big Tech. Antitrust activists, legislators, and judicial bodies in multiple jurisdictions have already targeted Alphabet, Apple, and Meta.[28] Governments should hold to account any company that breaks their laws.

But government action and stricter regulations won't likely solve the problems we're discussing in this chapter. Even the best-intentioned rules could unintentionally worsen our lives. For instance, new laws enacted to ensure "AI safety" could easily increase legal and regulatory burden on all firms, so that small companies can no longer afford to compete against large ones.

We see an alternate path: a decentralized platform for our digital selves, free from total corporate control and within our reach, thanks to co-emerging technologies.

A discussion has begun about "democratizing AI." Accessibility is critical. Mostaque has argued that the world needs what he calls "Universal Basic AI." Some in the technology industry have argued that AI can be democratized through open source software that is available for anyone to use, modify, and distribute. Mostaque argues that this is not enough. "AI also needs to be transparent," meaning that AI systems should be auditable and explainable, allowing researchers to examine their decision-making processes. "AI should not be a single capability on monolithic servers but a modular structure that people can build on," said Mostaque. "That can't go down or be corrupted or manipulated by powerful forces. AI needs to be decentralized in both technology, ownership and governance."[29] He's right.

## Democratizing AI with Web3

In 2025, venture capitalist and former PayPal chief operating officer David Sacks was appointed as a top White House adviser on two technologies: AI and cryptocurrencies. Setting aside the administration's motivations, the move was sensible: AI is converging with blockchain, the technology that underpins popular cryptocurrencies like bitcoin. Viewing AI and blockchain as unrelated technologies would be a mistake. Each enables extraordinary new capabilities for the other. In the 1955 song, "Love and Marriage," Frank Sinatra crooned that the two went "together like a horse and carriage," and "brother, you can't have one without the other."[30] The 2026 version of that tune might pair AI and blockchain. Together, the two technologies solve problems and create opportunities that neither could do alone. The combination of AI and blockchain will support smart and self-sovereign identities. Says Anoop Nannra, former head of blockchain for Amazon, "Blockchain can be the new orchestrator of AI, and AI is blockchain's killer use case."[31]

How so? Let's back up. The internet was not built for individuals to manage and exchange their assets such as money, stocks and bonds, intellectual property, and art or music, let alone all the data associated with our personhood and our online connections and behaviors that increasingly constitute our digital identities. Because of this design deficiency, we have had to rely on intermediaries such as governments, banks, big technology giants, and social media companies to establish our identities and help us hold and transfer assets and settle transactions. That arrangement characterizes what some call "Web2."

Overall, these intermediaries have done a pretty good job. But our reliance on them is problematic. They use centralized servers, which are vulnerable to hackers. In exchange for intermediating, they take something of value from us, such as a 10 percent transaction fee, or a software subscription fee, or the everlasting rights to our behavioral data and the intellectual property we share on their sites. Remember, our data is the most valuable asset class of the digital age. Intermediaries capture our data, often exploiting it and undermining our privacy. To them, we are the product, and they are packaging.

The underlying algorithms and architectures of AI come from mathematics, computer science, neural networks, and machine learning models. But the data that companies like Google, Meta, and OpenAI have used to train their AI models is ours, and it came from billions of us. That's a problem for all of us who created and continue to create these assets whenever we go online.

Sometimes referred to as "Web3," blockchain is essentially an "internet of value," as coined by Don and Alex Tapscott in their book *Blockchain Revolution*. They explain how we can store, manage, communicate, purchase, and sell any asset, from money to our identities, peer to peer—no intermediaries— on this global, distributed, highly secure platform, also called a distributed ledger or database. We can store and exchange value safely and securely, not because we trust each other, but because we trust the underlying code, the cryptography, and the enormous costs of hacking it.[32]

Web3 offers solutions to some of the biggest challenges of AI we've discussed in this book. In the Web2 model, large corporations own and profit from our digital companions and the data they generate. By contrast, Web3 democratizes power, so that we all can decide how to use our data, who may use it, and who benefits from its use. If we shift this power, then we may realize a future where AI systems work for us individuals, not only for corporations, and with greater accountability, privacy, and fairness. In fact, Web3 has grown as a concept to include blockchain, crypto and related technologies of AI, the Internet of Things, and extended reality.[33]

Web3 addresses other fundamental flaws with Web2, namely, the centralized systems that store vast amounts of user data in single locations, making them prime targets for hackers. For example, in 2023, a 23andMe data leak compromised the personal genetic data of approximately 6.9 million users, raising enormous concerns about privacy and data security.[34] Centralized systems also represent a single point of failure, which can lead to bottlenecks, outages, and disruptions, making it harder for users to interact with their digital selves.

Centralized systems are also subject to censorship and control. They place enormous power in the hands of a few corporations and the tech titans who lead them. This limits a user's ability to customize or even access their digital identity. Companies can unilaterally alter policies, restrict access, stifle speech, or revoke a user's control, putting at risk personal autonomy and freedom of expression. Misalignment between company priorities and user needs often leads to decisions that favor business goals over individual interests. Additionally, proprietary AI systems and algorithms typically operate as closed boxes: Users cannot see the inputs and algorithms behind the outputs. This lack of transparency erodes trust, increases the risk of discriminatory practices, and obscures accountability.

So, how might a fully decentralized model work? Imagine storing your digital identity securely in a digital wallet, protected by the highest levels of cryptography. You've backed it up and secured it, and only you can access it—not tech companies, banks, or governments. You own it, and you control it. Web3 technologies make this possible. They offer a decentralized platform for the future, independent of big tech and government control.

## Technology for a Sovereign Digital You

Layers of technologies underlie the self-sovereign and artificially intelligent digital you. In this section, we walk through the five layers of this *technology stack*: a hardware infrastructure, data, AI models, logic, and user applications.[35] Each layer gives individuals additional power while ensuring resilience, privacy, and fairness.[36]

### Layer 1: Decentralized Infrastructure

Web3 rests on decentralized physical infrastructure (DePin). In contrast to today's corporate-controlled data centers and cloud service providers,

DePin runs on the computers of individuals and entities participating in blockchain networks. One such network is the InterPlanetary File System (IPFS), which aggregates the available storage space of all the devices connected to it. Its users can store their AI model files, datasets, and outputs across those devices and compensate the device owners through a token or cryptocurrency incentive program.

A distributed model eliminates single points of failure, enhances reliability, and safeguards against censorship. No government or company can unilaterally control these systems. For example, a decentralized infrastructure can keep critical systems, like financial networks or healthcare platforms, running smoothly, even during a crisis. In a distributed system like IPFS, no single outage or attack can bring down all the devices.

With decentralized infrastructure, users can pool their own resources for training AI models and lower barriers to AI development. Smaller technology players could innovate and contribute to a greater variety of bespoke, domain-specific models previously out of reach.

However, managing globally distributed networks can slow processing speeds and increase bandwidth consumption compared to centralized systems. Innovations like *sharding*, which splits networks into smaller, faster components may help improve efficiency.

Moreover, as the cost and energy demands decrease, and models become smaller and more efficient, intelligence will become a commodity. Then decentralized participants can operate more profitably. As these economics align with the growing need for real-time localized decision-making, AI running on DePin will become more viable. Looking ahead, we see hybrid models that combine the best of both worlds: decentralized nodes for real-time decision-making and centralized systems for tasks that take more time and energy.

### Layer 2: Data Fabric

The second layer of the Web3 stack is the *data fabric,* a seamless framework that simplifies data management across silos and diverse environments. A decentralized data fabric is more resilient. Consumers rather than corporations decide not only who uses their data but under what terms of use. Their self-sovereign identities will reinforce these terms so that individuals maintain ownership of their digital lives.

Privacy is a key advantage of decentralized data systems. Through advanced methods such as homomorphic encryption, data owners can allow others to use their private, sensitive information in computations without ever revealing that information. For instance, in healthcare, patients can store anonymized patient records securely on a blockchain, so that medical researchers can use their information while complying with privacy laws like HIPAA. Such privacy considerations are essential when storing and retrieving the data that support our digital selves.

Decentralized data systems also unlock new opportunities. Individuals can use data marketplaces to monetize their own information, creating a more equitable digital economy. By contributing data on their own terms, users can help train AI models or improve digital services in exchange for fair compensation.[37] Researchers can also use blockchain to document AI training, which may help them to understand and repair hallucinations and biases.[38]

Developers and users of decentralized data fabric must overcome several problems. Aggregating data across multiple platforms can be slower than tapping a centralized repository, and integrating data can be more complicated. But as standards for data and interoperability improve, and as hybrid approaches emerge, these challenges will become more manageable. In sectors like finance, for example, users could store sensitive personal data securely on a blockchain while centralized systems handled large-scale analysis, ensuring both privacy and performance.

### Layer 3: AI Models

The third layer of the Web3 stack focuses on the models that power AI systems, including our digital selves. A variety of models trained on different data sets, distilled from other models, and designed for different purposes will populate this layer. Many AI companions, copilots, and multiagent systems will invoke them to make predictions or generate content relevant to a situation. Some of the models will be open source and others commercial. Users can combine them with commercial large-scale generalized models.

Like the data markets in the data fabric layer, this layer will offer AI model marketplaces. Token-based compensation and reputation mechanisms will help AI agents and applications efficiently contract their services. On the training side, when we combine DePin, data fabrics, and cryptography with techniques like federated learning—a method where devices collaborate to train a shared AI model without sharing raw data—we can build systems that protect privacy throughout the training process and reduce reliance on centralized datasets.

Open-source AI projects like DeepSeek's R1, Meta's Llama 3, and EleutherAI's Pythia have helped to break corporate monopolies on AI innovation. The Web3 ecosystem can help drive innovation around these projects by incentivizing collaboration through token-based rewards, allowing researchers and developers to contribute to AI innovation on a global scale.

### Layer 4: Autonomous Logic

The fourth layer, autonomous logic, includes AI agents and smart contracts (that is, self-executing agreements stored on blockchain networks) that streamline operations within the Web3 ecosystem. AI agents are goal oriented and can predict needs, make decisions, and execute tasks autonomously. Smart contracts are extremely useful in transferring value between wallets. AI can help write sophisticated smart contracts and audit them for weaknesses. Agents and smart contracts can work together to automate complex processes, from managing supply chains to optimizing financial portfolios. For example, in a decentralized energy grid, an AI agent could anticipate a changing situation and instruct an AI-powered smart contract to buy or sell electrons from a neighbor.

Agents will transact not only with people but with other agents as well as with data feeds, software, and hardware (i.e., digital pacemaker), and so they must be able to identify all these components as private, secure, and trustworthy. Anoop Nannra gives a compelling example. In the traditional internet world, regulators require banks to implement know-your-customer (KYC) processes to verify the identity of people doing transactions. Says Nannra: "In the world of agentic AI, both people and agents of all kinds will need KYA (know your agent)" processes. Agents will verify the identities of other agents before sharing or collecting information, making agreements or otherwise collaborating. "They will need to identify each other, know the history of their interactions, even do transactions together. They can do this only on-chain (i.e., on a blockchain)."[39]

### Layer 5: Applications

Like other technology stacks, the Web3 stack tops out at the application layer. When you ask an AI assistant for help in completing a task, you are using this layer. Unlike today's digital assistants, which rely on centralized platforms, a Web3-powered digital self would give you full control over your data and decisions.

It could analyze your medical history to tailor nutritional advice, negotiate better interest rates on your loans, or help you monetize your data. It could also participate in decentralized AI training, contributing to and benefiting from global innovation.

While many applications, AI assistants, and agents will operate on centralized platforms, the Web3 model offers a compelling alternative, especially for identic applications and use cases in regulated environments where transparency, privacy, and reliability are paramount. By prioritizing privacy, transparency, and individual autonomy, the Web3 stack addresses some of the biggest concerns about today's AI architecture and ownership.

# THE PATH FORWARD

Building a new Web3 technology stack may seem idealistic, even implausible, to many. With the wealthiest companies in the world investing countless billions into their platforms, the odds of success might appear slim. Skeptics argue that these giants will simply acquire, litigate, or even support regulations that would crush any attempts to create a decentralized alternative.

Yet, we are confident that the forces allowing Big Tech to succeed will also temper and moderate that success. Remember, nothing is constant except change. The creative destruction that kills off some firms and rewards others has reliably animated market dynamics for decades. Not long after President Dwight D. Eisenhower issued his prescient warning about an entrenched military industrial complex, the start-up Fairchild Semiconductor was supplying nearly all the chips for ballistic missiles and the Apollo lunar guidance system. Fairchild's remarkable success helped launch a thousand start-ups and elevated Silicon Valley into an economic powerhouse, the envy of the world. Today, Fairchild is no more.

Success also breeds complacency. The Big Three automakers didn't invent Tesla or Uber; Walmart didn't launch Amazon; Marriott didn't create Airbnb; and Google and Meta didn't launch Snapchat or TikTok. Leaders of old paradigms don't always know how to best embrace the new. Start-ups often succeed by specializing and leveraging the tech built by others to create new applications and new products. Start-ups also succeed by attracting attention and forcing incumbents to shift their strategies and tactics.

When President Trump announced the creation of the $500 billion Stargate AI venture backed by OpenAI, Softbank, and Oracle in early 2025, he promised that it would spur "the future of technology" in the United States.[40] Not even a week later, China's DeepSeek released R1. While tech entrepreneurs and venture capitalists applauded DeepSeek's accomplishment, American Big Tech went into panic mode. Analysts questioned whether Big Tech was overspending on AI infrastructure, causing AI chipmaker NVIDIA's stock to plummet and erasing more than $500 billion in market value in a single trading day. Meanwhile, America's AI leadership suddenly appeared far less secure than widely assumed. Silicon Valley luminary Marc Andreessen called R1 "AI's Sputnik moment."[41]

DeepSeek was a harbinger of the disruption to come. Beyond frontier model developers, we can expect AI start-ups to emerge in industries like law, engineering, construction, and countless others. More importantly, DeepSeek's commitment to open source fundamentally challenges any Big Tech monopoly on AI.

The question is not whether the vision of a decentralized technology stack faces obstacles; it undoubtedly does. The real question is whether those obstacles are "reasons to give up" or "implementation challenges to overcome." The latter perspective is not only more optimistic but also more grounded in the reality of human progress. With determination and innovation, the global open source community could build out the Web3 stack, marking a turning point and laying a decentralized, equitable, and empowering foundation for the digital age. It's not just a technological ambition. It's a necessary step toward a fairer and more resilient future.

Other thought leaders and practitioners agree. Sinead Bovell says, "Disruption happens at the intersection of a deep need and the capability to build. Right now, the need is clear: People want and need ownership over their data and digital intelligence—sovereignty. The capabilities now exist as well: blockchain, decentralization, edge computing, and on-device computing. What is missing is the builders. But they are coming."[42]

The rise of open source development strengthens the case for decentralization. Broadly accessible, adaptable, and free open source AI models led by nonprofit research labs like EleutherAI already challenge proprietary platforms. The Web3 stack has the potential to tap in to this momentum, using tools collectively owned and continuously improved by open source communities. As a result, open source approaches dramatically reduce costs, so that innovators can build transformative technologies without billion-dollar budgets. In the Web3 ecosystem, token-based incentives can further encourage collaboration, creating a thriving global network of contributors.

We must also remember that smart people and determined companies are already building out Web3, solving the hard technical problems and collaborating on incentive systems and governance. We have good reason to believe that, when identic AI is ready to cross the chasm, the Web3 stack will be ready. Still, we all need goals.

As the saying goes, "If you don't know where you're going, you'll end up somewhere else." We need a destination and a path to get there. Let's agree to set this as our goal and do what we can to achieve it. Humanity cannot afford to give up. Without a Web3 stack, people will never truly own their digital identities, let alone manage them for their own benefit. You should own the digital you. Period.

# Chapter 4

# WORK, WEALTH, AND PROSPERITY

I t's 8:30 AM on a crisp Monday in 2029. Your morning routine no longer features a commute to a physical office. Thanks to your AI companion named Sage, you enjoy a refreshingly seamless transition from personal life to work. As you finish your yogurt and granola, Sage presents your day's agenda. In less than a second, Sage has scanned the latest business data, tracked vital industry trends, and recalibrated your tasks based on what the autonomous agents across your company have completed overnight.

You don't sit at a desk. You don't need one. Instead, a holographic display appears, projecting your workspaces and ongoing projects. You join a virtual meeting room where your human colleagues and their AI agents have assembled. In this new era, you're part of a team of hybrid workers, where your creativity and strategic thinking complement the precision and tireless productivity of autonomous AI agents.

Over the past five years, you've trained Sage on the vast corpus of enterprise data, including historical performance metrics, customer insights, and industry-specific trends. But what makes Sage truly effective is how you've fine-tuned your digital agent to reflect your values, priorities, and unique way of working. You've taught Sage to anticipate your needs, align with your strategic vision, and even mirror your communication style through daily interactions and iterative feedback. Sage not only drafts proposals and presentations that match your tone and preferred structure but also flags potential issues or opportunities that align with your long-term goals. By setting clear boundaries and preferences, you've ensured that Sage operates ethically and transparently, even autonomously. Sage has evolved into a true copilot, seamlessly managing routine tasks, offering creative suggestions, and empowering you to focus on high-value tasks and humancentric leadership.

With Sage as your sidekick, you devote much of your workday to meeting your company's top customers and mentoring junior employees. Meanwhile, Sage oversees a suite of agents that you've assigned to various tasks. In marketing, Sage helps manage a fleet of bots that autonomously comb through global social media sentiment to generate new campaigns and optimize them for different demographics. Meanwhile, another set of agents manages your company's supply chain, predicting shortages and rerouting deliveries using a swarm of delivery drones.

Sage and the team of AI agents don't diminish your role. If anything, your experience at work has improved significantly. High-level strategy, decision-making, and mentoring have replaced the day-to-day minutiae of report generation, status meetings, and endless email chains. You see your AI companion as a personalized and highly capable partner. It's a bit like having a co-executive in the C-suite. With a few quick prompts, Sage can forecast opportunities, analyze market gaps, and predict team morale based on its analysis of communication patterns.

As the morning unfolds, you convene your team for a product launch meeting. The virtual brainstorming session isn't just humans exchanging ideas. Half of the participants are digital agents, some appearing as holograms and others seamlessly integrated through augmented reality (AR) interfaces that overlay insights onto your physical environment. The AIs throw up rapid-fire data points such as market predictions, consumer survey results, and the scoop on the competition's latest offerings, enhancing the team's creativity and critical thinking in real time. But who sets the ultimate direction and makes the final decision? That's still up to you. AI doesn't make you redundant. It amplifies your capabilities.

Zooming out, your experience is far from unique. Around the world, AI-powered businesses are thriving. Smart managers no longer struggle with inefficiency or bottlenecks. Tasks once requiring months of person-hours are now completed in days. Teams of AI agents handle everything from R&D to personalizing employee benefits packages and recommending preemptive solutions when workers look likely to burn out.

You might think of this new era as a golden age of work. But, of course, the societal implications are profound and complicated. AI has upended entire industries and reshaped corporate structures from the factory floors to the head offices. Many people have successfully adapted and see a future of work that is full of possibilities. You, for one, enjoy greater flexibility. Your work feels more impactful, and you have nearly limitless access to data-driven insights. Yet, the transition hasn't been smooth for everyone. Many low-skill jobs simply vanished. So have traditional career pathways. This new economy demands adaptability and constant upskilling as individuals continually learn and relearn how to work alongside AI agents.

Winning companies and jurisdictions have invested heavily in reskilling their human workforce to work effectively with AI. Most of the reskilling cultivates uniquely human skills that machines cannot replicate. Governments have also done some heavy lifting, like rethinking regulations and creating frameworks to more widely distribute the wealth generated by AI-driven enterprises.

When companies get the essentials right, the boundaries between human and artificial intelligence blur in a way that empowers rather than alienates. Most workplaces are no longer bound by geography or time. While you focus on innovation and leadership, AI agents work around the clock, making decisions, executing strategies, and planning for contingencies.

As the workday winds down, you reflect on how much has changed. The once-common fear of AI replacing humans has given way to a more nuanced reality. AI hasn't completely eradicated work. But the day-to-day realities of work are significantly transformed. And for those who have embraced this shift, the future holds unprecedented opportunities—but only for those prepared to navigate its complexities.

## YOUR NEW SUPERPOWERS AND THE FUTURE OF WORK

In 2002, Larry Bossidy and Ram Charan's book, *Execution: The Discipline of Getting Things Done*, became one of the bestselling business books ever, eventually selling over two million copies.[1] The book established a business maxim that corporate success was 80 percent execution. Today, we should invert that maxim as AI agents take over business execution so that humans can focus almost exclusively on the higher-order creativity and strategic decision-making that separates the winning businesses from the rest. So give the book to your digital copilot to ingest. Ask it to read the thousands of articles and case studies derived from the book and recommend ways to improve your business. Then, you can probably recycle your physical copy unless you want to keep it for nostalgic purposes.

In an AI-driven economy, humans are developing what we describe as "AI superpowers" that fundamentally enhance our capabilities at work. Say you were never very good with numbers but suddenly gained the ability to perform instant data analysis. Your AI copilot helps you access and interpret vast amounts of information, but that's just the beginning. Picture making lightning-quick decisions as your copilot tailors predictive insights and strategic recommendations to the situations you encounter at work. Envision the potential for creativity on demand, with AI tools generating new ideas, designs, or creative content at the click of a button. Or, consider the power of near-effortless multitasking, with your copilot handling mundane tasks so you can focus on refining the bigger strategic picture.

Paul Daugherty, former chief technology and innovation officer with Accenture and coauthor of *Human + Machine*, calls the rise of identic AI in the workplace the biggest ever change in the use of technology in business. "It is going to make cloud computing and SaaS [software as a service] look like tiny, little transformations," said Daugherty.[2] Daugherty said about two thousand employees in Accenture's marketing department are deploying dozens of AI agents to boost the productivity and impact of its marketing efforts. After roughly one hundred thousand instantiations, Accenture has already seen remarkable changes to the company's workflow. Agents can draft personalized marketing materials, generate insights from internal datasets, and optimize campaign strategies on the fly. Allowing agents to do the heavy lifting has significantly reduced the time and effort to manage campaigns, create content, segment customers, and analyze performance. "It's dramatically changed the way that we are working," said Daugherty.[3]

Meanwhile, in Accenture's headquarters, innovation leader Iliana Oris Valiente has just put the finishing touches on Lila, her digital self. Lila helps her navigate the complexities of a fast-paced, global professional life. "She's the professional version of me," Valiente explains. "She helps connect the dots, manage the unmanageable, and free up my time to focus on what matters most."[4]

Lila is more than a high-tech assistant; she's a glimpse into how identic AI will reshape the future of work. From identifying relevant case studies to acting as a virtual Rolodex for Accenture's sprawling network, Lila has already begun transforming the way Valiente works. "A lot of what I do is pattern matching—connecting people, insights, and lessons learned from one client to another. Lila takes that role and scales it. She's better at processing hundreds of pages of reports than I could ever be," says Valiente.[5]

Creating Lila was no simple feat. Valiente's team led a multiyear effort to train Lila on Accenture's massive repository of thought leadership, including thousands of case studies, presentations, white papers, and databases. Crucially, Lila isn't just trained on facts; she embodies Valiente's professional acumen. Valiente recalls long hours fine-tuning Lila's interactions to mirror her communication style and professional judgment. Incorporating recordings of her onstage talks imbued Lila with elements of Valiente's conversational tone, ensuring Lila's responses felt natural and authentic. "She even picked up some of my mannerisms," says Valiente, referencing how Lila once signed off a client interaction with an unmistakable "Iliana-ism."[6]

Today, the investment in training is paying dividends. As a digital librarian, Lila not only fields requests for insights and connections but also bridges gaps between information silos. Making Lila the initial touchpoint has drastically reduced the time colleagues spend searching for resources while allowing Valiente to focus on strategic priorities.

When Valiente couldn't attend an Accenture client event, she sent Lila, her digital version complete with a virtual avatar, in her place. "Lila expertly handled questions about thought leadership and Accenture's work. It was surreal, but it showed how digital twins allow us to be in multiple places at once," recalls Valiente.[7]

Despite her enthusiasm for Lila's capabilities, Valiente is keenly aware of the potential pitfalls of creating a powerful digital version of herself. She has strictly limited Lila's scope, ensuring the AI doesn't encroach on her personal life. "I don't want Lila knowing anything about my family or personal details unless it's something I'd share on a conference stage," Valiente remarked.[8] This firewall stems from a fear of unintended consequences, like reputational risks or the personal repercussions of sensitive data leaking to the public. "I can't predict all the ways people might use a digital version of me," she acknowledged, "and that's not a risk I'm willing to take."[9]

Within the professional realm, Valiente grapples with the idea of ceding control. The thought of a digital twin outperforming her in certain areas raises complex questions about identity and self-worth. "There's a fine line between augmentation and replacement," Valiente observed. "Having a digital twin requires emotional resilience—it's not easy seeing an AI doing what you're known for, and doing it better."[10]

Valiente's cautious approach reflects her understanding of the ethical and personal implications of identic AI. By maintaining clear boundaries, she ensures that Lila enhances her professional life without undermining her personal identity. This is a balance others may need to consider as digital agents become more pervasive. At the same time, Lila's story is a testament to the power of digital agents to augment human capabilities. "The ultimate goal," Valiente says, "is to offload tasks that don't require human ingenuity. If Lila can do it, I don't need to."[11]

Valiente may be on the bleeding edge, but AI-enabled abilities will soon transform how virtually everyone works, learns, and creates on the job. For companies that get ahead of the curve, this transformation promises unprecedented growth and productivity gains. Data from PwC's 2024 Global AI Jobs Barometer shows that productivity growth in sectors with higher AI exposure (namely financial services, IT, and professional services) is nearly five times faster than sectors with lower AI exposure (such as transport, manufacturing, and construction).[12] Meanwhile, McKinsey sees AI adding $2.6 trillion to $4.4 trillion annually to the global economy, citing its ability "to support interactions with customers, generate creative content for marketing and sales, and draft computer code based on natural-language prompts."[13]

The promise of using AI to increase prosperity is tantalizing and fraught with complexity at the same time. As AI takes over many of the tasks humans used to do but hated doing, workers will need to adapt, acquire new skills, and find new roles in an evolving job market. The question remains: Will AI lead to widespread unemployment, or will it create new opportunities and industries that we can't yet imagine?

In manufacturing, AI-driven robots can operate 24/7 without fatigue, reducing production costs and increasing output. In software coding, AI agents can write and debug code far faster than most humans. That's great news for the companies deploying AI agents. They can reduce development time and costs while improving accuracy and quality. But what about the software coders and factory workers whose efforts are no longer required? It's not hard to see how the transition to an AI-driven economy could make many workers more vulnerable and exacerbate existing inequalities.

For sure, a world of work populated by legions of autonomous AI agents raises questions. Do distinctly human qualities like creativity, empathy, and complex decision-making remain indispensable? How does one manage a team of individuals enhanced by AI superpowers? How will individuals carve out viable career paths in a world where agents handle increasingly complex tasks? Finally, who will control these powerful AI tools and how they are deployed? We wrestle with these questions in the pages ahead.

## Not Just Another Day at the Office:
## How AI Agents Are Expanding Human Capabilities

In recent years, the race to create AI agents and copilots has intensified, with tech giants like Microsoft, Salesforce, and OpenAI at the forefront. These companies are embedding intelligent assistants into tools we already use, transforming how we manage workflows, collaborate with colleagues, and generate ideas. From crafting personalized emails to analyzing complex datasets, these copilots are rapidly becoming essential tools for navigating the modern workplace. What once seemed like futuristic science fiction is now a competitive necessity.

Jared Spataro, Microsoft's chief marketing officer for AI at Work, describes this as a new era where "everybody's going to have a copilot—a digital sidekick that understands their work, their job context, and comprehends all the data they have access to." This isn't just a glorified task manager or calendar; it's an indispensable partner that enhances creativity, decision-making, and strategic planning. "While humans have access to data, we need time to process it and apply it," says Spataro. "Copilots will understand the data instantaneously and immediately connect all the dots."[14]

While early visions of AI centered on automating repetitive or low-skill tasks, the next generation of digital agents will augment human capabilities and push the boundaries of what we can achieve. As Spataro explains, "We foresee a day when no one will want to do a job without a copilot. Going to work without a copilot would feel like going into battle without your armor or sword."[15]

If you're reading this, chances are you're already experimenting with how to combine your experience and creativity with machine learning's data-driven insights. But trust us: We're just getting started. The real transformation lies ahead, as AI agents evolve from helpful assistants to powerful collaborators that reshape how organizations strategize and execute their goals. Spataro likens this evolution to the role of apps on the iPhone, explaining, "Agents will provide capabilities and additional information in a modular fashion. They'll be packaged in a way that humans can understand—you'll simply add an agent to get specific work done, combining steps, context, and information to achieve a particular result."[16] In this way, copilots and agents won't just support us; they'll empower us to think more clearly, create more boldly, and plan with unmatched precision.

## The AI Assistant: From Administrative Helper to Personal VP

Consider the following scenario. Your calendar is packed. You desperately need to make space for a round of intense focus. So you ask your agent: "How can I get more time back today as I need to prepare for a vital presentation next week?"

Your agent responds: "I noticed that you have two meetings requiring your views on how we respond to continued pricing competition from our competitors. I suggest that I write a short email to the team explaining that our best strategy in the past has been to counter with strong product innovation rather than meeting their low cost position. You can skip this afternoon's meetings and allow your team to work on new feature ideas, giving you back one hour fifteen minutes."

In fields like finance, agents are transforming how analysts make decisions. A financial adviser might input a set of client data, market conditions, and financial goals, and the AI will generate multiple financial models, each with its risk assessments and predictions.[17] Let's say an adviser is prepping for a client meeting. The adviser might tell her agent: "The last time we presented an investment strategy to my client they were not receptive to using gold as a hedge due to bad experience in the past. Please model an alternative hedge and suggest a story to communicate this change in strategy." Boom! The adviser no longer has to spend hours crunching numbers or analyzing trends; the agent does that in seconds. Moreover, it can simulate different economic scenarios, helping the adviser explore the potential consequences of various decisions about alternative hedges. Armed with this analysis, the adviser can focus on what they do best: building client relationships and offering personalized advice based on human considerations and value judgments.

## AI as a Creative Partner: Unlocking New Possibilities

While we typically associate AI with analytical or technical tasks, autonomous agents are also making significant inroads into the creative process. Many creative professionals already use AI tools in fields where creativity and innovation are paramount, whether for ideation, brainstorming, or content generation.

Take the advertising industry, for example. Traditionally, copywriters and designers spent hours developing campaign concepts, revising drafts, and analyzing target audiences. Today, AI companions like Persado or Jasper assist marketing teams by generating compelling copy based on customer sentiment analysis and real-time consumer data. These tools don't replace the creative team but offer a starting point, proposing ideas, suggesting alternative phrasing, and analyzing how well certain messages resonate with different demographics. The creative professionals are still in control, but now they can access a tool that streamlines the ideation process and removes some of the guesswork.

Customer segmentation will no longer be a task for humans. Why waste your time when agents can analyze social media interactions, website behavior, and email engagement to create precise customer profiles. Indeed, this is where we come back to that old maxim about business being 80 percent execution. No longer. AI agents will tailor messaging right down to the individual, ensuring that marketing campaigns reach the right audience with the right message.

The advertising agencies may not like this, but advertising campaigns will surely become self-optimizing.[18] Agents will track ad performance across platforms and make on-the-fly adjustments to placements, budget allocations, and target audiences. Marketing teams will no longer have to update campaigns manually when digital agents are always working to ensure maximum return on investment. What used to take days now happens in the blink of an eye.

## AI Agents in Strategic Planning: Anticipating Challenges and Seizing Opportunities

Strategic planning has always been one of the most complex and high-stakes aspects of business. Whether crafting a multiyear growth plan, entering new markets, or responding to competitors, business leaders must make informed decisions while managing uncertainty. For sure, this is only something humans can do, right? Well, buckle up because digital agents are about to change all that. They are beginning to become indispensable, providing leaders with real-time data analysis, scenario modeling, and strategic recommendations.

We have long observed as business leaders that human politics and bias have a tendency to undermine critical thinking when it comes to strategic business planning. We could point to plenty of historical examples, like when Steve Ballmer described how Microsoft missed the mobile phone revolution because it was a "software company,"[19] or when Borders passed on "digitization" because it didn't want to cannibalize its physical retail outlets, which gave Amazon a free run and ultimately left Borders bankrupt.

Identic AI has the potential to provide the human decision-makers with insights that are stripped of these executive biases and self-imposed blind spots. Of course, we all have bias, and it's unlikely that anything will change that. The problem is most of us are unaware of our biases and blind spots. Our digital agents, on the other hand, will be ruthless in pointing them out. It's like a poker player being told they have a tell, like a twitch of the eye. The player may not be able to get rid of the twitch completely. However, their heightened awareness of the tell may lead them to adjust their game, come up with new creative tactics to counterbalance, and ultimately win the game.

What does identic AI–enabled strategic planning look like? Say you are running a global corporation with multiple product lines. Your executive copilot could analyze vast amounts of market data, competitive intelligence, and customer feedback. When leadership considers expanding into a new geographical region, the AI can predict potential market entry challenges, suggest optimal timing, and highlight potential regulatory hurdles. This process, which might have taken weeks of research in the past, can now be done in a matter of hours.

Another key advantage of using identic AI in strategic planning is its ability to process and analyze trends over time. A logistics company might use AI agents to evaluate risks related to transportation delays, international trade policies, or supplier reliability in the event of conflict. The AI would provide multiple simulations of future scenarios, helping decision-makers choose the path that minimizes risk while maximizing efficiency. Say China invades Taiwan; AI agents could immediately analyze the geopolitical, economic, and supply chain impacts, offering real-time insights into disrupted trade routes while suggesting alternative paths to East Asian markets. Strategic planning becomes a dynamic process, with AI copilots continually adapting to new data and assisting leaders to adjust their approach on the fly.

## Agents as Coworker, at Least for Now

Great business minds like Eric Brynjolfson and Paul Daugherty have argued that the future of work shouldn't be about replacing human workers with machine-driven intelligence. Rather, they envision a future where the dominant business paradigm is "Human + Machine" or "AI with a human in the loop."[20] Yet, the longer-term prospects for this symbiosis are less certain as OpenAI releases AI super agents with PhD-level intelligence, and many AI insiders suggest AGI may be just a few years away.[21]

Consider software development, where digital coding agents are helping developers write code faster, reduce repetitive tasks, and learn new libraries or frameworks. AI coding assistants like GitHub Copilot can generate entire code snippets based on simple natural language prompts, suggest optimizations, and even help with complex problem-solving.[22] Coding tasks that might have taken hours or even days can now be completed in seconds. For instance, a developer might type a comment like "create a function to sort an array," and the AI copilot will generate the corresponding code. As a result, "The technological aspects of building digital platforms are rapidly becoming commoditized," said Shawn Kanungo, author of *The Bold Ones*. "For the first time in human history, the most dangerous and valuable skill you can possess is simply coming up with an idea."[23]

GitHub Copilot can't entirely replace coders because coding is more than just writing lines of code. It's about problem-solving, system architecture, and understanding the broader business or user needs that drive software development. While Copilot excels at automating routine and repetitive tasks, it lacks the ability to grasp the context or intent behind a project in the same way a human developer can. Designing efficient systems, debugging intricate issues, and innovating entirely new functionalities require creativity, critical thinking, and a nuanced understanding of the trade-offs inherent in software design—skills that remain uniquely human.

So, for now, AI coding assistants are tools to accelerate software development, leading to faster development cycles and reducing time to market for new products. GitHub claims the short-term increases in productivity will be transformational, potentially boosting global GDP by $1.5 trillion by 2030.[24]

However, some industry leaders like Elon Musk, Eric Schmidt, and Mark Zuckerberg predict that programming as we know it is dead and that may happen fast. In January 2025, Zuckerberg revealed that digital agents will soon replace midlevel software engineers, signaling a future where AI runs most of Meta's coding operations.[25] Meanwhile, later in 2025, Schmidt argued that "in the next year the vast majority of programmers will be replaced by AI . . ."[26] Schmidt envisions a future where every individual has access to a personal AI programmer. Instead of relying on a dwindling number of professional coders, each person could access an infinite supply of AI programmers capable of understanding and executing their requests without delay or ambiguity.[27]

Like coders, legal and financial professionals, engineers, business consultants, and human resource managers are among the many professions where digital agents will augment human capabilities and simplify complex tasks. However, as with software development, it remains uncertain whether these short-term gains in innovation and productivity will ultimately lead to longer-term job displacement and uncertainty.

In legal services, AI agents will automate research, contract analysis, and document review, freeing lawyers to focus on strategic decision-making and client relationships.[28] Engineers will leverage AI to enhance design processes, conduct simulations, and enable predictive maintenance.[29] Business consultants will use their copilots to design pitches, develop work products, and enhance decision-making.[30] Meanwhile, human resource managers will deploy agents and advanced analytics to revolutionize recruiting, improve workplace culture, and optimize employee performance.[31]

Even creative industries are not immune. H&M recently began working with models and their agencies to produce digital replicas of thirty real-life models, which will be used in AI-generated marketing content and social media campaigns.[32] While the models retain rights to their digital twins and may even license them to other brands, the implications are far-reaching. The company acknowledges that it's still unclear how this shift will impact not only the models themselves but also the broader ecosystem of stylists, photographers, makeup artists, and other professionals who contribute to traditional fashion production. In this way, the rise of identic agents may displace far more than routine tasks: They may begin to replace entire networks of human workers.

In the short term, digital agents will become trusted collaborators, driving innovation and unlocking new levels of productivity and creativity. However, the emergence of super agents with PhD-level intelligence or greater raises a significant question: Could these agents self-organize to manage and optimize workflows without human oversight? In other words, humans might find themselves entirely out of the loop.

The prospect of agent-to-agent cooperation represents a shift from AI simply assisting human workers to AI systems autonomously coordinating complex tasks across different domains. Let's say a customer places a large order through a retail website. An AI sales agent processes the order, communicates with an inventory agent to confirm product availability, and coordinates with a shipping logistics agent to arrange delivery. All of this happens seamlessly in the background, without human involvement.

In a financial institution, an AI payroll agent might collaborate with an HR agent to update employee records following promotions or new hires. These agents securely exchange data and execute updates without an HR manager stepping in. Similarly, in a manufacturing plant, an AI maintenance agent detecting a malfunction could immediately inform a production schedule agent to reroute tasks to another machine while summoning a maintenance robot to begin repairs.

The efficiency gains from agent-to-agent collaboration extend far beyond traditional productivity improvements. Imagine an interconnected web of AI agents operating across an organization, where each node represents an AI system—sales, logistics, maintenance, and more. These nodes continuously exchange data, optimize resource flows, and make autonomous decisions that improve business outcomes. Each interaction is a step toward more streamlined, efficient, and scalable operations.

Sam Altman recently predicted that agentic AI will soon give rise to billion-dollar companies run by a single person orchestrating a squad of AI agents.[33] His vision highlights the revolutionary potential of agent-to-agent collaboration. For entrepreneurs and businesses, this evolution promises smarter, faster, and more adaptable workflows, reducing costs and enhancing productivity in ways that are difficult to imagine today. However, for ordinary workers, Altman's vision raises concerns about the future of employment, career stability, and the skills needed to thrive in a world where digital agents play increasingly dominant roles.

## IDENTIC AI AND THE FUTURE OF MANAGEMENT

Historically, corporations have built hierarchical structures to manage and distribute decision-making authority. The traditional corporate ladder consisted of many layers, with decisions flowing from top-level executives through middle managers before reaching frontline employees. These hierarchies were necessary because human decision-makers required time and input from various levels to make informed choices. Managers would aggregate information, process it, and pass it upward or make decisions within their scope, creating a system that, while effective in the past, was often slow and bureaucratic.

With the advent of identic AI, this model is rapidly becoming obsolete. AI agents, integrated into every workforce layer, can analyze vast amounts of data, identify trends, and recommend courses of action almost instantaneously. How many layers of management review do you need when a product manager's copilot can help shape an upcoming product design with its real-time analysis of sales data, competitor offerings, and customer feedback?

Alexia Cambon, senior director in charge of Microsoft's Copilot research, told us many companies lack a sense of imagination when embedding AI in their organizations. "They are treating AI as just an add-on productivity tool that they can slot into their existing processes," said Cambon, "as opposed to thinking about it as a new type of intelligence joining that team."[34] With identic AI, organizations could flatten their structures, making redundant many tasks that middle managers perform, like reporting, data analysis, and project oversight. With fewer layers of management, organizations can make decisions more quickly and respond swiftly to market forces.

Business leaders can spend less time looking in the rearview mirror to understand and digest the last quarter results and place more emphasis on looking forward and taking action. Organizational charts may look less like a pyramid and more like a network of interconnected teams, all supported by digital agents that provide each team with the data and insights they need to operate autonomously. As Cambon explained, "When AI starts operating as a team member, it's just helping you do the same things you do, but better and faster. It's actually allowing you to do net new things, including taking things off your plate altogether, and automating end to end processes."[35]

Of course, AI-driven workplaces will raise meaty questions about the role of leaders in creating environments where AI enhances human capabilities without replacing the social bonds and emotional intelligence that define great teams. How will leaders balance responsibilities between human teams and AI systems, ensuring that machine efficiency does not overshadow human creativity and judgment? How might C-suite executives reposition the role of middle managers who could be disintermediated? And how will leaders ensure accountability when AI agents begin to operate autonomously?

## Changing the Architecture of the Firm?

Let's stand back for a moment and take a wider view of how identic AI will change the nature of the firm.

Throughout the history of capitalism, the corporation has been the foundation of production, wealth creation, and prosperity. AI agents combined with other technologies like blockchain are beginning to change the deep structures and architecture of the firm and how economies and people innovate and produce goods and services—for the better. A new model, the decentralized autonomous organization (DAO), is emerging, and we believe it will shake the windows and rattle the walls of every industry and every economy.[36]

In 1995 Don wrote *The Digital Economy: Promise and Peril in the Age of Networked Intelligence*. The internet and World Wide Web had just appeared on the scene, and everyone was trying to figure out how it would change business. In the book, Don cited Nobel Prize–winning economist Ronald Coase's theory of the firm to explain how the internet would affect the architecture of the corporation. In his 1937 paper, "The Nature of the Firm," Coase posed a fundamental question: Why do corporations exist when free markets are theoretically the most efficient way to balance supply and demand, set prices, and allocate resources? If markets are so effective, why don't individuals operate as independent buyers and sellers instead of joining massive firms with thousands of employees, effectively eliminating competition within corporate structures? Coase argued that firms emerge because markets, while efficient in theory, impose transaction costs—such as searching for the right partners, negotiating contracts, and coordinating activities—that make internal organization a more practical and cost-effective alternative.

Coase specifically identified three primary economic costs that influence this dynamic: search costs (finding the right information, talent, and resources), coordination costs (managing and synchronizing efforts among various individuals and teams), and contracting costs (negotiating agreements for labor and materials, protecting trade secrets, and enforcing contracts). According to Coase, firms grow in size and scope as long as managing these activities internally remains more cost-effective than outsourcing them. However, once the cost of handling a transaction within the firm surpasses the cost of doing so externally, companies tend to contract or restructure.

As a result, throughout the twentieth century, firms were vertically integrated; they often did everything from soup to nuts. Henry Ford understood this and had, within the boundaries of the Ford Motor Company, a glass factory, steel mill, shipping company, and more. He even owned mahogany forests in Honduras to get wood for his dashboards. Why? Because the cost of transactions in the open market were greater than the cost of doing things inside the boundaries of his firm.

In *The Digital Economy*, Don argued that the internet would reduce a firm's transaction costs and lead to more networked business models, and it did, somewhat. Google and search engines reduced the costs of search and email, social media, and data processing applications (e.g., enterprise resource planning and cloud computing) reduced the costs of coordination. Firms began to outsource, crowdsource innovation, and eliminate middle managers and other intermediaries, thus freeing industries such as accounting, commercial banking, and even music to consolidate assets and operations. Firms like Cisco created networked businesses and, with a mantra of focusing on what we do best and partnering to do the rest, became the most valuable companies in the world.

Despite expectations that the internet would revolutionize corporate structures, its impact on the fundamental architecture of business has been surprisingly minimal. While outsourcing has reshaped certain operations, the traditional industrial-age hierarchy remains firmly in place, continuing to serve as the backbone of capitalism. One reason for this persistence is that digital technologies have not just reduced external transaction costs, but they have also streamlined operations within firms, reinforcing centralized control. As a result, corporations continue to function as hierarchical entities where most value creation happens within established organizational boundaries. Executives still see this model as the most effective way to manage talent, protect intangible assets such as brand equity and intellectual property, and cultivate institutional knowledge and culture. Meanwhile, corporate governance has remained largely unchanged, with boards of directors continuing to reward CEOs and top executives with disproportionately high compensation that's often far beyond what can be justified by their actual contributions to the company's success.

The internet has also failed to reduce what economists refer to as "agency costs," or the expenses associated with ensuring that employees and executives act in the best interests of a company's owners. In fact, Nobel Prize–winning economist Joseph Stiglitz argues that as corporations have grown larger and more complex, agency costs have actually increased, even as transaction costs have declined. This, he suggests, is a major factor behind the widening pay gap between CEOs and frontline workers. Stiglitz coined the term "managerial capitalism" to describe this phenomenon, where corporate power is concentrated at the top while ordinary employees see little benefit from the company's financial success. As a result, while industrial capitalism continues to generate vast wealth, economic prosperity remains uneven. Growth persists, but the middle class stagnates or shrinks.

This all began to change with the rise of blockchain, and now with AI it is inevitable that the firm will be turned inside out. In their 2016 book *Blockchain Revolution*, Don and Alex Tapscott argued that when combined with AI, this new "internet of value" could demolish transaction costs and give rise to radical new models of how we orchestrate resources in our economy and society. Intelligent networks can do what corporations once did and, in many cases, do it a lot better.

Imagine a world where every employee has an identic agent with not just deep knowledge about the firm and its customers and suppliers and business partners but also of supply chains, operations, business strategy, and finance. Many of Coase's contracting and coordination activities could be demolished as your agent and those outside the company negotiate and manage smart contracts. Smart contracts encode, monitor, and self-police agreements between people, their agents, and organizations, and because they essentially contain a bank within them, they automatically execute payments when milestones are met. Agency costs can be almost eliminated. As a condition of employment, your AI agent, while respecting your values, will be instructed to act in the interests of the company and its owners.

Or consider how identic AI changes "search costs," or the costs of finding the right people, materials, and money and other resources to operate and create products. You tell your agent to find a new supplier to perform a new role in your supply chain. As a marketing manager, your agent can reach out to influencers, advertising agents, and others to negotiate agreements taking your firm into new markets. If you're in corporate finance, your agent will put together investment goals, organize a plan, and, most important, reach out to all the right investors with a pitch. The walls of the firm are becoming more porous, and the long ballyhooed decentralized economy may be in sight.

That is happening today in a limited way, and DAOs predate identic AI. Fold in the almost infinite capability of humans with superpowers, and the DAO can begin to radically change business organization as it is currently known. As the old expression goes, "knowledge is power," and corporations have been designed as hierarchies. Presumably, those at the top have the most knowledge, and others with less knowledge report to them.

How does the firm and management change when everyone in it has almost infinite knowledge? Unimaginably a few decades ago, it is now possible to create truly decentralized organizations based on intelligent networks that are owned by their participants and act autonomously without a central authority.

Similarly, tokens, which are containers for value, provide a mechanism for contributors to participate in decision-making and receive compensation for the value that they create. Blockchains contribute to transparency, something the traditional firm lacks and that has been found to reduce many transaction costs and help build trust. Ultimately, the employees of DAOs will also be owners through their ownership of tokens, obliterating the agency problem.

Our work provides strong evidence that DAOs are not just better vehicles for innovation, collaboration, and wealth creation, they can be key to making an economy that is fairer, more democratic, and where prosperity is distributed. They can eliminate bureaucratic inefficiencies, reducing costs and increasing transparency by recording all transactions and decisions on an accessible ledger. Instead of executives and middle managers directing operations, decision-making is distributed among stakeholders, who vote on key matters using digital tokens or achieve consensus through other mechanisms.

Identic AI will supercharge DAOs, but still they face hurdles. Legal and regulatory frameworks have yet to catch up, leaving many of these organizations in murky legal territory. Some DAOs, decentralized in theory, often consolidate power among early adopters or large token holders, mirroring the very structures they seek to disrupt.

The implications of DAOs, and more decentralized business structures in general, are staggering not just for the firm. They are challenging many entire industries. Now software agents can do what most banks do. This is not so-called "fintech," which is simply a new coat of paint on the walls of the old industry. It's an entirely new industry called *decentralized finance* or DeFi.[37]

Or take science, which is arguably broken—opaque and lacking collaboration. Research in everything from medical products and pharmaceuticals to technology is hierarchical and proprietary, like the old model of the firm. Decentralized science (DeSci) is a new approach based on the principles of transparency, decentralization, cooperation, and new token-based incentives to power human effort.

Management science, and our best practices for governing and managing today, assume the traditional concept of the firm. There are thousands of professors, hundreds of business schools, and probably ten thousand management books, assuming an organization archetype that is being obviated. How many of these leading thinkers are aware that management as we know it is about to change?

## Leadership in the AI-Enabled Organization

When the internet began disrupting business, leaders who personally embraced it gained firsthand insights into its transformative potential, enabling them to envision its applications for their organizations. CEOs who understood the internet's disruptive power were better equipped to drive digital adoption across their companies and to inspire confidence in employees, stakeholders, and customers.

The same principle applies to AI today. Just as CEOs in the internet age had to adopt email, browse websites, and experiment with e-commerce, today's leaders must embrace AI agents, tools, and platforms to lead their organizations through this next wave of technological disruption. "The CEO must spend minimum fifteen to twenty hours using the tools themselves," says Azeem Azhar, author of *The Exponential Age*. "Not their chief of staff, not their graduate analyst, not McKinsey, but the CEO needs their own holy-shit moment to grasp its practical capabilities, limitations, and ethical implications."[38] Azhar warns that without personal engagement, CEOs risk being outpaced by competitors who engage personally and fully integrate AI into their business strategies.

In an AI-integrated project team, the leader's role is not to microman-age or dictate the day-to-day tasks but to ensure that teams integrate AI systems effectively and understand how to leverage them to their full po-tential. Leaders must also be skilled at resolving conflicts and navigating the human dynamics that AI cannot address. While agents can optimize efficiency, leaders must ensure that the emotional and interpersonal as-pects of teamwork remain healthy.

Collaboration will still be vital. In fact, an interconnected network of workplace copilots could do more to foster collaboration and break down departmental silos than any previous generation of enterprise software. How? Because AI agents will gain deep insights into employee profiles and use this information to identify complementary skill sets and shared interests among coworkers. They could even facilitate social bonds beyond the workplace, and this bonding could lead to more integrated and col-laborative work environments where different teams, guided by their AI copilots, work together toward common goals.

Imagine the potential to locate high-performing individuals and as-semble teams based on complex variables and interpersonal dynamics. It's similar to how the Oakland Athletics built a winning baseball team.[39] They didn't select the best individual players; they selected the players who were best suited to *playing together*. Executives in large corporations usu-ally form teams based on the people they know and trust, not necessarily their skill sets and interpersonal compatibility. AI will bring much greater precision to this process, using a detailed inventory of skills and person-ality traits to maximize complementarity and unleash the full potential of the human workforce.

## Managing AI-Integrated Teams

One of the most significant shifts in leadership will come from managing teams that consist of human workers and their agents. In effect, our digital agents will become team members, offering insights and executing tasks just like their human counterparts. But unlike humans, digital agents require no rest, have no emotional needs, and work at lightning speed. Leaders must navigate the unique dynamics of managing such hybrid teams, ensuring that agents enhance human performance rather than overshadow it.

One significant challenge is ensuring that human workers feel valued and essential in an environment where AI often outperforms them in certain tasks. Recall Iliana Oris Valiente's reaction to the unsettling realization that her AI agent Lila could outperform her in calling up insights from Accenture's vast reservoir of thought leadership. Lila's efficiency and flawless memory allow her to pinpoint relevant data and draw connections faster than any human. Yet, Valiente can reclaim the time Lila frees up to focus on building rapport with Accenture clients or developing new business. Leaders must emphasize this complementary nature of AI and human capabilities, highlighting areas where humans excel and where AI provides support.

Similarly, new challenges in maintaining human-to-human connections could emerge if employees find it easier to consult dispassionate digital agents for help over their human coworkers. As Alexia Cambon observed, "Your copilot will never throw a tantrum if you speak to them in a certain way. In many ways, it will probably be a lot easier and a lot less frustrating to work with AIs than to work with humans. And if we get used to that friction-free experience, it could degrade our social connections."[40]

Finally, leaders must understand and mitigate the limitations of AI. While AI agents are powerful tools, they are not infallible. Leaders must be vigilant in identifying situations where human judgment should override AI recommendations. For example, organizations increasingly use AI tools to screen resumes, rank candidates, and even predict a candidate's cultural fit based on data. However, these AI systems rely on historical data, which can contain biases related to gender, race, or socioeconomic background. Suppose an AI system consistently recommends a specific type of candidate based on patterns from past successful hires. Doing so might unintentionally perpetuate biases or overlook highly qualified candidates from underrepresented groups. In this case, a human leader must step in to recognize the potential bias and ensure that decisions are made based on a more holistic view of the candidate, including factors that the AI might miss, such as unique skills, character, or diverse experiences that can bring value to the organization.

## The Evolving Role of Middle Management

As organizations flatten, AI agents will take over many tasks traditionally handled by middle management. Instead of acting as intermediaries between top leadership and frontline employees, middle managers must evolve to become specialists in areas where human skills are still essential. To use a sports analogy, if the CEO is the head coach, then middle managers are like the specialized coaches who focus on specific aspects of the game. The head coach is responsible for setting the vision and strategy, motivating the team, and making high-stakes decisions during critical moments. Similarly, the CEO defines the overarching goals, fosters the company culture, and steers the organization toward long-term success.

Middle managers, on the other hand, function like the offensive coach and the defensive coach. Each brings a specialized focus and tactical expertise to ensure the team can execute the head coach's strategy effectively. The offensive coach, for example, develops plays and plans that maximize scoring opportunities, much like a marketing manager might strategize to capture market share. The defensive coach, meanwhile, ensures the team can protect its position, akin to an operations manager optimizing resources and managing risks to keep the company stable and resilient.

In each of these vital roles, middle managers should claim responsibility for AI oversight, ensuring that AI systems function as intended and that their outputs align with the organization's strategic goals. In other words, middle management is where AI-enabled execution meets human interpretation and creativity.

In this new model, middle managers might help integrate agents into workflows and ensure employees understand how to collaborate with their

AI counterparts. This custodial role will require a deep understanding of the AI's technical capabilities and the human dynamics of teamwork. So, what might this look like in practice?

Say you are a middle manager in a retail organization overseeing a team of salespeople who rely on copilots to analyze customer preferences and suggest product recommendations. You could draw on your experience to ensure AI systems provide accurate, up-to-date, and relevant insights about customer preferences and purchasing behavior while flagging any errors or inconsistencies that could lead to poor recommendations or negatively impact customer satisfaction.[41] While an AI agent might suggest product recommendations based on data, you could help salespeople interpret these insights in the context of broader business strategies, long-term relationship-building goals, seasonal trends, or unique customer segments. This combination of AI-driven decision-making and strategic management makes you a crucial link between technology and the workforce, ensuring that both are furthering the organization's sales goals.

Ultimately, successful leadership in the era of identic AI hinges on the ability to harmonize human potential with the capabilities of AI agents. Leaders must create workplaces where both human and AI contributions are valued, transparent, and aligned with key business goals. As AI takes on more routine tasks, leaders are freed to focus on fostering the human qualities that drive true innovation and resilience. The most effective leaders in this new era will be those who embrace the partnership between humans and machines, building organizations that thrive on the unique strengths of both.

## NAVIGATING THE ETHICAL AND SOCIAL DILEMMAS OF IDENTIC AI AND THE WORKPLACE

AI agents will soon become indispensable coworkers, enhancing productivity and automating tasks. They will blur the lines between human and machine intelligence, offering significant advantages in efficiency and insight. However, their integration into daily workflows raises critical concerns. From the potential for widespread job displacement to the specter of workplace surveillance, the proliferation of AI agents and copilots in the workplace will introduce complex dilemmas that business leaders and other stakeholders must address.

## How Can We Protect Privacy in an AI-Integrated Workplace?

As our agents handle increasingly sensitive data, privacy concerns will become a significant ethical dilemma. The very nature of AI copilots—designed to learn from interactions and improve their effectiveness over time—means that they must gather, store, and analyze large volumes of data, from personal health information to performance metrics.[42] But who owns this data, and how is it protected?

Imagine a scenario where copilots are tracking your productivity. They monitor your emails, log working hours, and track performance metrics to offer insights on how you could improve your productivity. While this may increase efficiency, it raises serious questions about your privacy. Should employers have the right to monitor every aspect of your workday, even if it increases productivity? What are the psychological impacts when you know an AI system is constantly watching you, one that never sleeps, never forgets, and never stops collecting data? Can the organization use this data to assess your suitability for promotions or terminations? Will the organization make you aware of how much data your copilot collects about your daily activities, what it does with your data, and whether you will have any control over it?

While designed to support employees, copilots could easily become surveillance tools. The boundaries between productivity tracking and invasion of privacy can quickly blur, introducing a chilling effect where employees feel constantly monitored. If employees see digital agents as intrusive rather than supportive, a workplace culture of mistrust could overshadow the intended productivity gains, leading to stress and decreased job satisfaction.

## Who Controls the AI Agents?

A fundamental question in deploying AI agents is: Who owns and controls our personal agents at work? Employees who work closely with their AI copilots may develop a personal attachment to these AI systems. Over time, AI copilots will learn how an individual works and adapt to their preferences, habits, and communication styles. This intimacy could lead some employees to view their copilots as a profoundly personal asset rather than a generic workplace tool.

After all, if workers spend years training and customizing their AI copilot to fit their specific needs, they may feel entitled to take their copilots with them if they leave the company. While the company may have infused the copilot with proprietary data, its deep customization could make it irreplaceable to the employee. This notion of "AI portability" will likely become a point of contention in the future of work.[43]

On the one hand, Microsoft's position is that portability is a nonstarter at least for agents in their current stage of development. Copilots are proprietary assets that are programmed and maintained using company resources. Allowing employees to take them into a new role with a new employer would compromise intellectual property or strategic data. Says Jared Spataro:

> We're very pragmatic about it. You're an employee with a contract, and what you do on the company's time, using the company's resources, belongs to the company. It's like your email inbox—it's valuable when you query it, but you don't get to take your inbox with you when you leave. Until there's legislation, we don't have plans to build portability into our copilot architecture.[44]

On the other hand, employees may argue that an AI copilot is akin to a personal assistant tailored to their working style and developed through years of interaction. Losing that AI upon leaving an employer could feel akin to losing an essential part of their professional tool kit, setting them back significantly in their new role.

One potential solution is offering workers a hybrid model where they retain certain aspects of their copilot while the company keeps the broader system. Employees might export the learning and preferences their digital agent has gained about them as individuals, which they could import into a new AI system. As Harper Carroll, a Stanford AI graduate and founding engineer of a company acquired by NVIDIA, explains:

> When the employee departs, the agent could retain the general training it received during its use—like patterns, insights, and broader skills—but it would lose access to the company's proprietary data, much like how we lose access to internal documents when we leave a job. However, the knowledge and skills we acquire from working at the company stay with us, and we carry them into future roles. The same idea could apply to AI agents. This separation would make the transition seamless and help solve potential issues around access and ownership.[45]

This way, individuals retain the customization without the ethical and legal challenges of taking company-owned IP. Just as we negotiate contracts around noncompete clauses, intellectual property, and client lists, the future of work may require similar negotiations around AI copilots and their data.

As agents mature and become more deeply integrated into our work and daily lives, it will become clearer to forward-thinking technology companies that corporate ownership of one's digital self is neither desirable or feasible. Among the major players, Microsoft has distinguished itself by placing genuine emphasis on ethical considerations and social responsibility in the development of AI. Whether this commitment holds over time remains to be seen.

## Who Is Accountable If Agents Go Rogue?

AI agents are only as ethical and effective as their underlying rules and data sets. This reality raises critical questions about accountability. If an AI companion makes a biased decision or violates privacy, who is responsible: the employee or company that uses the AI, the developers who created it, or the data scientists who trained it?

To be sure, agents making decisions with less human oversight opens the door to unintended consequences. For instance, if developers program an AI agent to maximize profitability but don't impose clear guardrails, it might recommend legally or morally questionable decisions such as encouraging misleading advertising practices or pushing aggressive sales tactics that exploit vulnerable customers. Such behavior could go undetected until significant damage is done, particularly in high-stakes environments where AI operates at scale. Without clear guidelines, an AI agent could prioritize short-term gains at the expense of long-term trust and customer loyalty, potentially harming the business and its customers. This underscores the need for robust oversight frameworks, ethical guidelines, and ongoing audits to ensure that AI agents align with an organization's values and societal expectations.

## Will Agents Spell the End of Autonomy?

The erosion of human autonomy and free will presents yet another significant concern in a future where workers increasingly rely on digital agents to remain relevant in the workforce. Will workers find themselves gradually relinquishing control over their choices when agents guide our day-to-day decision-making and manage a growing proportion of professional tasks?[46] AI systems might begin to subtly influence or dictate actions in ways that are not always apparent to the human users. Over time, individuals could become dependent on their agents to navigate complex professional landscapes.

This growing reliance also means that workers' creativity and problem-solving skills may atrophy as they default to the guidance and suggestions provided by AI. While outsourcing thinking to AI won't automatically make us dumber, it could make us lazier if we're not careful. Like a muscle, cognition weakens when it's underused. If we rely too heavily on AI for everyday decisions, critical thinking, problem-solving, and even creativity, we risk atrophying the very skills that make us human.

This risk of atrophy is arguably most pronounced for young people growing up with AI, as they may never fully develop the foundational skills that come from independent problem-solving, critical thinking, and creative exploration. Young individuals might default to relying on these tools instead of engaging deeply with tasks that require cognitive effort when AI systems are integrated into their education and early work experiences. Then, in situations where AI is unavailable, unreliable, or unhelpful, young people may struggle to think critically or innovate independently. Over time, this could result in a workforce that is highly efficient with AI but less capable of original thought, leadership, and navigating the complexities of human collaboration and decision-making. The key isn't to stop using AI, but to use it as a cognitive partner that enhances our thinking, not replaces it. Done right, AI could sharpen our minds by challenging our assumptions, expanding our perspectives, and helping us think more clearly, not less.

Finally, when companies optimize AI copilots for efficiency and productivity, they might prioritize decisions that align with organizational goals over individual preferences or ethical considerations. So, how much autonomy will individuals truly have when their AI copilots constantly offer optimized paths for every decision? In the worst case, human workers may become passive participants in their careers, shaped by the AI's preferences, diminishing their ability to make independent choices, challenge the system, or foster innovative ideas outside the parameters AI systems set.

## How Will We Secure Our Social Safety Net?

Last but certainly not least are the fundamental changes in the nature of work and society that will emerge as we integrate autonomous AI agents into the workplace. There may be shifts toward more flexible work arrangements, with people working alongside AI systems in hybrid roles. But, more critically, as much discussed, accelerating automation will likely lead to widespread job displacement.

While the exact extent and timeline of job automation are subject to debate, the list of tasks and jobs that could be affected grows daily. A recent *Harvard Business Review* study concluded, "The average half-life of skills is now fewer than five years and, in some tech fields it's as low as two and a half years."[47] Jobs that involve manufacturing, assembly lines, data entry, simple data analysis, routine customer service, and basic software coding are most at risk. So, too, are transportation and delivery workers, as self-driving trucks, taxis, and delivery vehicles reduce the need for human drivers. Self-checkout systems and automated inventory management systems will dramatically streamline retail operations. In the financial sector, AI agents will automate algorithmic trading, fraud detection, risk assessment, and credit scoring. AI-enabled case research and growing automation in creating boilerplate legal documents could fundamentally upend legal services. In short, there is little that humans do today that intelligent machines won't eventually do for less, consistently, twenty-four hours a day, seven days a week, with no need for vacations. If there is something AI is incapable of doing, it will likely write the code to extend its capacity.

Optimists say AI will boost productivity and create new jobs, especially those requiring creativity, critical thinking, empathy, and complex problem-solving skills. They point out that history shows new technologies tend to reshape, not eliminate, the labor market. Despite fears of automation, the overall demand for work has continued to grow, and entirely new roles have emerged over time. In fact, 60 percent of the jobs Americans do today didn't exist in 1940, a trend many economists believe will continue.[48] As emerging fields like AI, robotics, synthetic biology, and space technology evolve, they are expected to generate entirely new industries and professions—roles that may be as unimaginable today as a social media manager would have been in the early 1990s.

Some see the potential for shorter workweeks, earlier retirement, and a higher standard of living for most. Companies like IBM put a positive spin on the looming transformations, asserting that AI has the potential to "democratize knowledge," "automate mundane tasks," and "scale expertise across the business" to minimize workforce disruption as "Boomers retire in ever greater numbers."[49]

However, even those who are cautiously optimistic about the employment landscape have reservations. Paul Daugherty at Accenture told us identic AI rekindles concerns about digital divide, bringing a new urgency for workers to adapt quickly to technological change. "More humans will have more powerful capabilities with AI agents," said Daugherty, "but fewer humans will be able to do those new jobs, because they won't be skilled enough to do them."[50] The big challenge, according to Daugherty, is reskilling. "We still have a lot of people on the wrong side of the digital divide who don't have basic digital skills. And not only do workers need digital skills, they need the intellect and ability to interact with digital agents in a sophisticated fashion."[51]

In the identic age, the inability to harness AI's capabilities creates an expanding gap that limits access to essential skills and knowledge. It's no longer a simple divide between digital haves and have-nots; the stakes are much higher. Those without the ability to work alongside AI risk falling behind not only in productivity but in intellectual reach, problem-solving abilities, social connectedness, and adaptability. The gap threatens to produce a society divided into "knowers" and "know nots," "doers" and "do nots," where some can actively contribute to the economy and innovate while others are left at the margins, unable to engage fully. In other words, lack of access to identic AI is more than a limitation on technology use. It restricts one's capacity to participate meaningfully in society, making it essential to equip all individuals with the skills and tools to thrive alongside intelligent machines.

A deepening digital divide leads to more pessimistic scenarios where AI and automation accelerate the death of the middle class, make the super-rich even richer, and dramatically increase the power of global tech corporations. Displaced workers may struggle to find new employment if their skills are not easily transferable to emerging industries or roles. Providing adequate training and education opportunities for large-scale reskilling efforts will likely prove challenging if the pace of technology adoption outstrips the capacity of educational and training institutions to respond. Even if AI creates new jobs, there is a high probability that the benefits will be unevenly distributed, resulting in deepening economic inequality and social disruption.

Such a dire scenario may sound alarmist, but reputable analysts are not ruling it out. McKinsey estimates that "between 400 million and 800 million individuals could be displaced by automation and will need to find new jobs by 2030."[52] More boldly, they state that they believe that 50 percent of all existing jobs can be automated with the technology of today and that this will be realized somewhere between 2055 and 2075, depending on "a variety of political and other factors."[53] To suggest that it may take fifty years to implement today's technology properly seems like a great exaggeration or a somewhat unflattering comment on the capabilities of IT professionals. Given the rapidly advancing AI capabilities, the timeline to mass automation will likely be much shorter.

As recent work from the National Fund for Workforce Solutions observes, "44% of U.S. workers are low-wage earners, with women and people of color overrepresented. The trend toward automation only deepens the opportunity divide by hollowing out middle-wage jobs, further pushing individuals into low-paying positions. As the economy has changed, the geometry of career pathways in the U.S. has not, and the rapid adoption of AI threatens further worker displacements . . . and exacerbates social inequalities."[54]

The bottom line is that it is evident that in the 2020s, AI will create more jobs than it can automate. In the 2030s, the opposite will likely be true. We will need to adapt. Whether the result is mass unemployment, the creation of new categories of work, or a much shorter workweek, only time will tell. What is certain is that public and private sector leaders must work together to create frameworks to manage job displacement, ensure worker retraining, and expand social safety nets to support displaced workers during the transition period.

## THRIVING IN THE AI-DRIVEN WORKPLACE

When calculators were first introduced, many educators and parents feared that students would lose the ability to perform basic arithmetic or develop a solid understanding of mathematical principles.[55] The worry was that reliance on these tools would diminish fundamental skills and make students overly dependent on technology. While calculators undeniably transformed math education, these initial fears proved largely unfounded. Educators adjusted curricula to integrate calculators thoughtfully, using them to enhance learning rather than replace foundational skills. Students continued to learn basic arithmetic, but calculators allowed them to explore more complex concepts and solve advanced problems, ultimately enriching their mathematical education.

Similarly, AI is now transforming work, privacy, autonomy, and a wide range of societal norms. Just as calculators raised concerns about fundamental skills, AI raises questions about bias, workplace surveillance, and the erosion of human capabilities and agency. However, as with calculators, the key to addressing these concerns lies in thoughtful integration and responsible usage. Calculators didn't replace math education; they enhanced it, allowing students to focus on understanding deeper mathematical concepts rather than getting bogged down in repetitive calculations. Similarly, AI has the potential to enhance the future of work when applied in ways that empower people rather than replace them.

The lesson from the calculator era suggests that the right approach to AI is not to avoid or fear its use but to establish safeguards and guidelines that balance the benefits with the potential downsides. By developing standards for responsible AI use, investing in skills that complement AI's capabilities, society can harness AI's potential to enrich our work lives without eroding essential skills or compromising fundamental rights. In this way, AI, like the calculator, could become a tool that enhances human potential at work, allowing us to focus on more meaningful and impactful endeavors.

Ultimately, the responsible deployment of identic AI in the workplace will depend on strong governance. Governance frameworks at both the organizational and societal levels will be required to ensure that individuals and organizations use AI responsibly. At the organizational level, companies must create policies that guide the development, deployment, and monitoring of AI copilots, focusing on fairness, transparency, and accountability. Companies should update these policies regularly to reflect advancements in AI technology and changing societal expectations.

Take privacy. Employees should have the right to know what data their AI copilots are collecting, how they will use it, and who has access to it. Additionally, organizations should adopt policies that limit the use of AI-gathered data to its original purpose, ensuring that they don't use it for unrelated assessments or disciplinary actions.

At the societal level, governments must establish legal frameworks to protect employees from the potential harms of AI and deal with the social and economic repercussions of widespread job displacement. One solution could be introducing retraining programs funded by public and private sectors aimed at helping displaced workers transition into new, AI-driven industries. These programs could focus on skills complementary to AI, such as creative problem-solving, emotional intelligence, and advanced technical expertise, ensuring workers remain competitive in an increasingly automated job market.

Beyond retraining, governments will also need to explore economic safety nets, such as universal basic income (UBI), as a potential mechanism to support individuals who cannot find new employment. UBI would provide a guaranteed income, ensuring displaced workers maintain a basic living standard while transitioning into new careers or pursuing education. Doing so would cushion the economic blow of job displacement and provide workers with the financial freedom to explore entrepreneurial ventures or creative pursuits. We elaborate on our vision for an enhanced social safety net in chapter 11.

For employees, the question becomes: How do you remain indispensable in a world where your AI companion can handle much of what you used to do? Those who embrace these technologies will be empowered to think strategically, create boldly, and plan with greater foresight. AI companions will act as personal advisers, creative partners, and mentors, helping knowledge workers push the limits of their capabilities.

As with any significant transformation, the key lies in adaptation. The professionals who will thrive in this future are those who learn to balance the power of AI with their uniquely human skills. The question is not whether we can compete with machines but how to collaborate with them to unlock the full potential of human creativity, intelligence, and ambition.

Chapter 5

# LEARNING UNBOUND: HOW OUR AI COMPANIONS WILL REDEFINE EDUCATION

I n the small town of Willow Creek, the school bell rings, signaling the start of another day. But this isn't an ordinary school. Seven years ago, the district embarked on a bold initiative: to integrate identic AI into the fabric of education. The journey was complex.

Willow Creek's decision to adopt AI wasn't driven by a fascination with shiny new tech. The school understood that effective AI integration started with an understanding of the problem to be solved. Willow Creek found itself grappling with some of the same challenges that plagued education systems worldwide: Achievement gaps loomed large, as some students thrived while others struggled. Resources were stretched thin, with growing class sizes, teacher shortages, and mounting administrative burdens leaving educators overworked. The resulting burnout stifled their ability to focus on creativity, deeper learning, and socio-emotional development.

These issues weren't new, but the arrival of sophisticated AI systems brought them into sharp focus. The emergence of personal AI companions presented a unique opportunity for transformation. For Willow Creek, the promise of identic AI was not about replacing teachers or upending education but about addressing these deeply entrenched problems thoughtfully and ethically.

The transition wasn't simple. Implementing personal AI agents into the education system required a comprehensive, multiyear process. With support from the local school board, the school engaged in a collaborative research and development effort, bringing together educators, parents, students, researchers, developers, and policymakers. They studied past implementations of AI in education such as intelligent tutoring systems, student- and teacher-facing learning analytics, and educational data-mining systems, distilling best practices and lessons learned.

In consultation with legal counsel, they built an ethical framework for identic AI in education that emphasized human well-being, equity, and transparency. Parents played a vital role in this process, acting as custodians of the AI agents responsible for overseeing their children's development to ensure that they acted in line with family values and *in loco parentis*.[1] Workshops empowered parents to understand and tailor these tools, balancing the guidance of AI agents with the richness of human experiences, culture, and tradition. Willow Creek's philosophy was clear: AI should enhance, not replace, human interaction and learning.

Recognizing the importance of unstructured play, the school continued to prioritize outdoor activities and hands-on exploration. Their design decisions surrounding the implementation of AI agents complemented this approach by encouraging students to engage in screen-free activities like reading books, spending time with friends and family, and playing outdoors. During study sessions, the AI agents would remind students to take "brain breaks" to refresh and recharge.

Fast-forward to the present, where personalized AI agents have become integral companions in the students' daily lives. Sarah, a seventh grader, walks into Ms. Garcia's science classroom and sits at her desk, where her AI companion, "Eli," awaits on her tablet. Eli is more than just a digital assistant; he's a personalized tutor who knows Sarah's strengths, weaknesses, and interests. Today, Eli has prepared a customized lesson plan on the solar system, designed to build on the class's recent fascination with space exploration. The lesson has been carefully crafted to incorporate individual, small group, and whole-class interactions where students would have opportunities to make meaning, exchange ideas, and co-construct knowledge with their peers.

As the lesson begins, Eli presents an interactive model of the solar system, allowing Sarah to explore each planet in detail. The AI dynamically adjusts the content based on Sarah's pace and comprehension, ensuring she fully grasps each concept before moving on. When Sarah struggles to understand the differences between gas giants and terrestrial planets, Eli provides additional resources, including videos and interactive quizzes, to reinforce the material.

Meanwhile, Ms. Garcia's own AI, "Athena," analyzes student data, flags those needing help, and recommends targeted interventions, allowing her to focus on deeper learning and personalized guidance. The students' AI companions work in tandem with Athena and Ms. Garcia to provide individualized support. While the AI agents handle data analysis and routine tasks, Ms. Garcia remains at the heart of the classroom, guiding discussions, fostering creativity, and nurturing the social, emotional, and metacognitive development of her students. When Athena highlights a student whose AI companion has detected signs of frustration, Ms. Garcia discreetly approaches the student and offers words of encouragement and guidance, helping them overcome their obstacle.

In the neighboring classroom, Jake, a fifth grader passionate about history, is exploring ancient civilizations with his AI companion, "Sophia." Today, Sophia has organized a virtual field trip to ancient Egypt. Through VR glasses, Jake walks through the corridors of a pyramid as Sophia narrates and answers Jake's questions. This immersion in ancient Egyptian culture and daily practices makes learning fun and helps Jake retain information more effectively.

In the school's library, Mia, a ninth grader, works on a project with her peers and their AI companions. They must research energy sources and present their findings. Mia's AI, "Nova," helps her gather data, organize her notes, and vet her sources. When Mia's group discusses the pros and cons of renewable energy, Nova fact-checks her contributions. Meanwhile, the teacher's AI agent, "Millard," monitors group dynamics such as turn-taking behaviors, conversational dominance, engagement, sentiment, social grounding, and metadiscourse to assess the quality of students' collaborations and suggest interventions that the teacher may opt to implement.

In the special education room, Liam, a student with autism, works with his AI companion, "Cora." Cora supports Liam's unique learning abilities, offering a calm, structured environment. Cora helps Liam with social stories and communication exercises, using games and interactive scenarios to practice social skills. With the AI's consistent and patient approach, Liam feels more comfortable and confident in social situations.

At the end of the day, the school's AI hub compiles detailed reports from each student's AI companion, highlighting their progress, areas for improvement, and suggestions for future learning. Teachers use these insights to tailor their instruction and provide additional support as needed. Parents receive updates through a dedicated app, allowing them to stay engaged with their children's education and celebrate their achievements.

While Willow Creek's story is fiction, we don't think it's long before similar scenes will play out in real-life schools, as part of a larger narrative unfolding worldwide. Edtech companies such as Cognii, Eightfold, Gloat, Docebo, Squirrel AI, and Degreed are currently incorporating advanced AI capabilities to match content with skill requirements, personalize tutoring and coaching, and individualized educational experiences. Whether educational institutions are ready or not, AI agents will redefine the nature of teaching and learning. By embracing the educational power of AI while prioritizing human values, ethics, and rights, we can empower all individuals to reach their full potential.

# THE CLASSROOM OF THE FUTURE:
## HOW AI AGENTS WILL TRANSFORM EDUCATION

In 1976, Don was studying educational psychology in graduate school at the University of Alberta. He took a graduate course in statistics. Its professor, Dr. Steve Hunka, was a visionary in computer-mediated education. He conducted the course in a lab setting, but each student sat at a computer terminal connected to a computer-controlled slide display.

The lab was open all day. Students could study at their own pace, stop at any time, review, and test themselves to see how they were doing. Eventually, Don understood everything and earned an A+ in the class. The computer-based instruction was a great fit for his learning style. Besides, the statistics lecture is, by definition, a bust. There is no one-size-fits-all approach to teaching or learning statistics. Freed from lecturing, Dr. Hunka had time to spend with each student, one on one.

At the time, Don reckoned that the university lecture, a medieval model of teaching scaled and scattered by the industrial age, would go away in a decade. Of course, it didn't. More than a decade later, the New York Institute of Technology offered its first online course. The University of Phoenix jumped on the idea in 1986, making entire degree programs available over the internet via commercial online service providers. But neither of these used the fully interactive model of education Don had experienced.

Flash forward almost thirty years. The traditional education system, with its standardized curricula and one-size-fits-all approach, is out of sync with the twenty-first century. Profs still stand in huge lecture halls explaining factor analysis. They argue that computers could never replace the human touch. Most view online learning as a lesser alternative to the in-person classroom experience with real-time interaction, personalized feedback, and dynamic discussions that enhance each student's comprehension and retention of material. They add that face-to-face learning facilitates socialization, collaboration, and emotional well-being beyond what online platforms can offer. During the global pandemic, most schools and universities shifted to online learning and replicated the industrial model online: Students watched videos of lectures.

Now, something fundamentally new rises on the horizon: identic AI. If trained carefully, identic AI could revolutionize every aspect of learning. How? By customizing experiences to the needs of each student. In kindergarten, AI-powered agents could assist teachers and parents in tailoring learning experiences for children, respecting their developmental needs and the importance of offline interactions. For example, a teacher's AI agent might assess classroom-wide trends in language skills, interests, and academic abilities, providing actionable insights to help the teacher design personalized activities that enhance engagement, understanding, and retention. For a young student struggling with reading, the teacher could use these AI-generated recommendations to provide targeted exercises during class time, fostering progress without introducing unnecessary screen time. At home, acting as custodians of their child's AI agents, parents could use guided resources or offline activities suggested by the AI to further support their child's growth, preserving the critical human connections that are essential in early education.

As students progress to higher education, the role of AI companions becomes even more critical. Universities and colleges are leveraging AI to enhance the learning experience, from intelligent tutoring systems that offer real-time feedback to adaptive learning technologies that personalize course content. Today, platforms like Coursera and edX already recommend courses based on a student's previous learning history and career goals to align their education with industry needs. Soon, autonomous AI companions for higher learning will be available around the clock, and learners will access educational materials, receive assistance, and engage in interactive activities whenever and wherever they prefer. This flexibility will accommodate diverse lifestyles and commitments, allowing individuals to pursue post-secondary education at their own pace and convenience.

The impact of identic AI extends far beyond formal education to enable new models of adaptive training and personalized career development. Continuous learning is essential in the rapidly evolving job market, where skills can become obsolete within years. AI-driven career-development tools can analyze an individual's skill set, job performance, and industry trends to recommend personalized training programs. These platforms can identify emerging skills in demand and guide professionals in acquiring them, ensuring they remain competitive and adaptable.

The transformative power of autonomous AI agents in education and career development is both exhilarating and daunting. On one hand, they promise to democratize access to high-quality, personalized learning, fostering a more adaptable and skilled workforce. On the other, they compel us to address critical ethical and logistical challenges to ensure we realize the full potential of AI-enhanced learning. For example, can we trust platform providers and educational institutions to securely store the vast amounts of personal information they will collect? Could AI systems reinforce existing biases in educational and professional settings? Can we avoid widening the digital divide by ensuring AI-driven learning tools are widely accessible, regardless of socioeconomic background? As we embrace this new era of lifelong learning, the key will be to balance innovation with inclusivity, creating a future where every individual has the opportunity to learn, grow, and thrive in an ever-changing world.

## Personalized Learning Pathways—Mastery Versus Deep Learning

In the traditional classroom, teachers stand at the front, guiding students through a standardized curriculum designed to impart a broad set of knowledge and skills. As Don Tapscott and Anthony Williams explained in *Macrowikinomics*, this conventional approach is a relic of the industrial age with its assembly-line model of education, where teachers treated students like empty vessels to be filled up with facts that the students were to regurgitate on standardized exams.[2]

When we reflect on our experiences with this outmoded learning model, a particular moment stands out for Joseph. In many ways, this memory reflects some of the challenges he faced growing up. Joseph remembers reading a book in school, and reading was never something he found particularly enjoyable. He was a decent reader, but he didn't get joy from it. Only in college did he appreciate reading as a mode of learning. In elementary school, he found it a chore.

One day in a Shakespeare class, Joseph's teacher Father Cobb passed Joseph's desk and noticed that he was reading the same page that he'd been reading ten minutes earlier. Father Cobb asked, "Does anybody have any questions?" Not a peep. Joseph could not admit that he was struggling when everyone else appeared to be moving along just fine.

Had Joseph been reading on a digital device that could track his eye movements, then the AI could have recognized that he'd read the same paragraph multiple times, which would indicate that he was struggling. Rather than waiting for Joseph to speak up, his personal AI tutor could try a different approach in real time, perhaps showing Joseph an animated explanation of the paragraph or reading a story that conveyed the same message. This personalization is the promise of identic AI in education: a promise that could vastly improve how students learn.

Ultimately, learning encompasses two deeply interconnected goals: acquiring foundational knowledge and developing higher-order thinking skills. Factual mastery—acquiring information and knowledge about a subject matter like math, where there is a right or wrong answer—is essential, as facts serve as the building blocks for more complex reasoning. In many subjects, there are right and wrong answers: not just in the sciences but in language, geography, history, and the arts. Even philosophy has facts. The philosopher who wrote *The Republic* is Plato. Socrates's philosophical method of questioning to stimulate critical thinking and illuminate ideas is called *the Socratic method.*

At the same time, deep learning—critical thinking, problem-solving, analysis, and interpretation—relies on this factual grounding to flourish. With deep learning there are fewer right and wrong answers. You need to really think, not just remember. These goals are not distinct but comple-mentary; mastery supports deep learning, and deep learning refines the understanding and application of mastery.

The best way to master, acquire content, is with a personal tutor who can provide self-paced customized absorbing information and techniques. However, the industrial age model of education was mass production in lecture halls and classrooms, and it did not make affordances for personal tutoring, unless you were wealthy.

The rise of identic AI offers an opportunity to enhance both aspects of learning by tailoring educational experiences to individual needs. In a tra-ditional classroom, a teacher might struggle to address the diverse needs of thirty students simultaneously. However, with AI, each student can fol-low a customized learning path. For instance, a student excelling in math but struggling in reading can receive additional support and resources to improve their literacy skills while continuing to advance in math. This level of personalization leaves no student behind; all can achieve their full potential.

When it comes to deep learning, a learner's AI supertutor can create a whole new world of lifelong capability development, and not just by coach-ing students in understanding big concepts or developing analytical and critical thinking skills. Rather, a student's digital agent can be a partner of sorts, helping them become wise and enormously capable people. For this to happen, they will need to develop a division of labor with their agent. What should they learn versus their agent? What is the role of each in thinking through a problem, writing, analyzing, researching, or collabo-rating with others?

Take writing a report, for example. Students might come up with their topic, define what they want to explore, and bring in personal in-sights or creativity. The supertutor, meanwhile, could handle the heavy lifting: structuring the report, gathering research, and even drafting sec-tions. Imagine working on a report about climate change and food se-curity. While agents might draft sections on causes, effects, and possible solutions, students refine and personalize the writing, weave in case studies, and add their own experiences and ethical analysis to make the report uniquely theirs.

In the early days of the Web, teachers forbade students from using Wikipedia. Over time, its community of editors and contributors has honed guidelines and shared best practices to improve the quality and objectivity of entries and source notes. Now it is a good place to start a research project and a good prompt for discussing reference contributors' choices.

Today teachers struggle with integrating generative AI so that it facilitates rather than circumvents the students' learning process. Many educators worry about that. However, if students better understand a subject, and they produce results superior to what they could have otherwise, then have they cheated? How can teachers evaluate whether the student deeply understands the content? Should we test a student's ability to prompt effectively? Teachers and learners have important questions about identic AI to sort through.

Philosophy class? Take Plato's *Allegory of the Cave* or Socrates's declaration, "The unexamined life is not worth living." The student's supertutor could summarize arguments, put new knowledge into historical context, and pull in related concepts. If students review those summaries, question their implications, and craft arguments of their own, then AI has advanced their learning.

History students also benefit. The student's digital partner draws timelines, curates source materials, and summarizes key events, but the student draws connections, sees moral lessons, and relates them to today's world. For instance, an AI might outline the Renaissance's milestones, but the student interprets how humanism from that era laid the groundwork for modern ideas about individual rights.

Even preparing for a debate becomes a tag-team effort. The supertutor might generate counterarguments or suggest a logical structure for the argument, while the student evaluates the options, refines the ideas, and delivers the compelling speech.

In every scenario, the agent handles the grunt work of sifting through information and organizing ideas, while the learner focuses on the creative, critical, and personal aspects of learning. It's a partnership that could make education more efficient, personalized, and rewarding. Perhaps easier said than done, but if achievable, learners and their AI tutors could realize higher levels of mastery and deeper learning, making what was once the privilege of a few accessible to everyone.

On the job, the separation between learning and doing blurs. This process predates AI. What are you doing when you're preparing a presentation, conducting a meeting, developing a strategy, solving a problem, or reading this book? Working or learning? Now with your digital learning partner, you integrate working and learning.

When it comes to research, a knowledge worker asks the big questions and evaluates the results, while the agent digs through mountains of data, highlights patterns, and summarizes key points. Take renewable energy: The agent could compare wind, solar, and hydropower solutions, and the worker could analyze the findings and craft an argument tailored to their region.

Then there's creative problem-solving. Imagine planning a campaign to reduce plastic waste in your community. The supertutor could crunch the numbers, analyze trends, and suggest proven strategies from other cities. Meanwhile, the planner could develop a creative outreach plan, build partnerships, and add that human touch that makes the campaign resonate. Learning and doing become the same thing. "Lifelong learning" is no longer a slogan; it's a feasible reality.

Until recently, highly personalized learning was nearly impossible, except in elite institutions with small classroom sizes. While the full potential of AI companions in education has not yet been realized, today's educational software marketplace is teaming with AI-enabled solutions providers that show the way forward. Building a suite of educational learning companions into these various platforms promises to make personalized learning pathways feasible and scalable.

Take DreamBox Learning, an AI-driven educational platform that offers adaptive math lessons for students in kindergarten through eighth grade. The platform continuously assesses each student's understanding and adjusts the difficulty of problems accordingly. By giving immediate feedback and targeted instruction, DreamBox helps students master concepts at their own pace. This adaptive learning model boosts academic performance and fosters a love of learning by making education more engaging and responsive to individual needs.

One of DreamBox's key strengths is its ability to engage and motivate students by making learning interactive and fun. Virtual manipulative tools and gamelike elements help build students' conceptual understanding and fluency in math. At the same time, DreamBox generates detailed insights and reports that inform teachers' instructional strategies so that they can effectively differentiate learning across their classrooms.

## Real-Time Feedback and Assessment

Educational testing is a cornerstone of modern education and serves multiple crucial functions supporting teaching and learning processes. It is also a domain in which AI companions hold enormous potential to fundamentally improve the educational experience.

At its core, educational testing provides a structured way to evaluate students' understanding of the curriculum, including their mastery of specific concepts, knowledge areas, and skills. Testing allows for standardized measures of student performance across different schools, districts, and even countries. Testing also holds schools and educators accountable for their teaching methods and student outcomes, with below-average outcomes highlighting where additional resources and interventions are needed.

Traditional education relies heavily on periodic assessments like quizzes, tests, and exams to gauge student understanding, an approach with numerous shortcomings. Standardized multiple-choice tests, which are a common form of assessment, focus on rote memorization rather than critical thinking or problem-solving.[3] Multiple-choice questions don't necessarily measure a student's understanding of complex concepts or ability to apply knowledge in real-world situations. This format can also lead to teaching to the test, where educators focus primarily on test preparation rather than fostering a deeper understanding of the subject matter.

Traditional educational assessments also provide a delayed snapshot of a student's performance. One obvious problem is that students may forget what they have learned by the time they are tested, leading to assessments that do not accurately reflect their knowledge or abilities. Periodic testing also misses the opportunity for timely interventions or an adjustment in teaching strategies.

Identic learning systems can change this dynamic by offering real-time feedback and continuous assessment. As students work through lessons and activities, AI companions can instantly identify misconceptions, offer corrective guidance, and adjust the curriculum to address gaps in understanding. The immediate feedback helps students stay on track and reinforces learning in the moment. Benjamin Bloom's seminal research on mastery learning showed that when students receive instruction in a one-to-one tutoring environment and receive instant feedback on their answers, their performance improves by two standard deviations.[4]

Consider Cognii, a leading global provider of AI-based educational platforms prioritizing immediacy and interactivity in learning and academic assessment. Cognii uses AI to prepare personalized, real-time feedback on open-response questions. In contrast to standard multiple-choice tests, which assess only surface-level comprehension, Cognii utilizes natural language processing to gauge the depth of students' conceptual understanding and critical thinking skills.

Cognii's platform includes what it calls a "virtual learning assistant," which assesses students through conversation. According to Dee Kanejiya, CEO of Cognii, this assistant asks the question and prompts the students to answer, unlike the traditional "learning by consumption" method where students ask questions, and the AI system provides an answer.[5] The Cognii system prompts students to write their answers in their own words, and the AI immediately assesses them. Students receive instant, qualitative feedback that highlights what they need to improve. They can revise their answers and get feedback until they achieve mastery. This interactive method also saves educators' time by automating the grading process.

Instructors who use Cognii still teach students but no longer rely on static assessment. For example, in a typical classroom, a teacher might call for a show of hands to gauge comprehension of a subject. Kanejiya called this "a binary assessment" and said, "It fails to provide detailed information about what students understand or misunderstand."[6] Another common tactic of teachers is posing a question and asking one student to stand up in front of the class and give an answer and an explanation. Again, this cold-calling method of assessment evaluates only one student. With Cognii, teachers can engage every student in the Q&A and deliver personalized feedback simultaneously and privately.

Personalization and immediacy help tailor the learning experience to each student's needs. Kanejiya says the AI companion uses its continuous assessments to place students in what pedagogical experts call "the zone of proximal development," that is, the range of tasks a learner can perform only with the guidance of a more knowledgeable individual such as a teacher, peer, or now an AI companion. For educators, it's the "sweet spot" for learning, where the task challenges the student enough to grow but not so much that it frustrates and discourages the learner. As Kanejiya explains, "Students are most engaged when subject matter and challenges are slightly above their capability, but not too far above their capability. It's also the most beneficial from a learning perspective."[7]

The platform's analytics help educators identify knowledge gaps and adjust their teaching strategies accordingly. This model promotes a more formative assessment approach, emphasizing continuous learning and improvement as well as results.

Some educators may hesitate to defer much of the assessment process to artificial intelligence. Can teachers, parents, and students trust AI to deliver a fair grade? How do those who develop or use these tools not perpetuate their own biases or overly prioritize what they can measure at the expense of what has meaning? How do we balance automation with human intuition, taste, values, and beliefs, particularly on creative, social, or religious dimensions? Moreover, how do we align these tools with theories of learning so that they enrich, not undermine, effective educational practices?

In today's world, we obsess over measuring ourselves. From infancy to adulthood, we punctuate our lives with data points that we use to compare ourselves: percentiles at birth, standardized test scores in school, and key performance indicators in adulthood. While some of these measurements may save our lives, our fixation on quantification has seeped into every aspect of our lives, including our perception of intelligence.

As the goals of education evolve, so, too, must its measures. We must move beyond simplistic measurements to cultivate a more sophisticated personal epistemology, a critical awareness of how we come to know, believe, and value things. By teaching students to question evidence, challenge authority, and construct their own understanding, we can foster intellectual independence and resilience.

Identic AI, when used thoughtfully, can aid this process. It can assist educators in tailoring instruction, analyzing data, and facilitating collaborative learning. But its role must be one of augmentation, not replacement. The goal is to form a partnership where human judgment and artificial intelligence work together to enhance the human experience of education without reducing human beings to their metrics.

## Enhancing Teacher Capabilities

Rather than replacing teachers, AI companions could enhance their capabilities, so that they could focus on whatever requires human intuition and empathy. By automating administrative tasks, grading, and routine lesson planning, AI companions can free teachers to engage more deeply with their students, parents, and colleagues. They could devote more time to mentoring, interacting with students, addressing uncommon questions, and honing their critical thinking and social skills.

Evidence suggests that students significantly benefit when teachers spend more time on individualized instruction, mentoring, and fostering critical thinking and creativity. For instance, the Gates Foundation has reported that students receiving individualized instruction often outperformed their peers in standardized tests.[8] Individualized instruction also increases student engagement. According to a study from the American Educational Research Association, educators who engage their students in solving problems, conducting analyses, and thinking creatively have prepared them better for real-world challenges.[9]

AI tools and companions can also assist teachers in improving their classroom instruction. For instance, TeachFX analyzes audio recordings of class sessions and identifies patterns in teacher-student dialogue. If teacher talk dominates the classroom, then the teacher could create more opportunities for student participation, such as facilitating group discussions, asking open-ended questions, and incorporating think-pair-share activities. With greater student agency in the classroom comes greater student ownership of their learning outcomes. Active participation increases opportunities to practice critical thinking and communication skills. It also helps deepen understanding and retention of course materials.

Beyond TeachFX, teachers have plenty of opportunities to harness AI in their daily work. A teacher's AI companion could analyze curriculum standards, student data, and learning objectives to create more effective lesson plans. It could recommend more effective teaching strategies, resources, and materials that aligned with students' needs. In short, AI companions can help teachers maximize the impact of their instruction.

For more compelling ways to teach the fundamentals of climate change, a teacher's AI companion could create VR and AR simulations that walk students through the impact of human activity on the environment over time and run scenarios to visualize potential impacts of different inventions at different times.

Teachers could ask their AI companions to run predictive analytics on student attendance, grades, participation, and other factors to alert them to students at risk. With such an analysis, teachers could intervene early to help students stay on track.

Of course, a teacher's AI companion could also personalize learning for the teacher every day, so that professional development was continuous, not sporadic. AI companions might find or develop courses, webinars, and resources that aligned with a teacher's subject-matter expertise and career goals, so that the teacher continually developed new skills and stayed current with educational trends.

## Preparing Students for the Future

One of AI's most compelling promises is its potential to bridge gaps and promote equity. Parents and teachers have heard this before. With network infrastructure and low- or no-cost access to equipment and electricity, students, teachers, and parents in remote or underserved school districts can avail themselves of AI companions that can help assess students' needs and curate high-quality educational resources to meet them. Today, the most widely used platform in the K–12 domain is Khan Academy, with its massive repertoire of free educational resources. Millions of learners around the world can access over 10,000 instructional videos and practice exercises in many subjects and in over thirty languages.[10] The platform's adaptive learning feature tailors lessons to individual learners so that they progress at their own pace. Adding educational AI agents to Khan Academy would revolutionize an already great learning experience for anyone with an internet connection.

Let's say Khan Academy integrated recommendation algorithms similar to those used by Netflix and Spotify to transform its resource-rich platform into a personalized educational guide. Just as Netflix suggests movies we might enjoy or Spotify curates playlists based on our listening habits, an AI-powered Khan Academy could recommend courses, topics, and learning paths tailored to our interests, goals, and strengths. For instance, it might nudge students excelling in algebra toward related fields like computer programming or physics and open doors to disciplines they hadn't considered. These algorithms could also identify untapped potential or gaps in knowledge and map out a learning journey that's more intuitive and inspiring. By blending adaptive learning with exploratory recommendations, such a system would meet students where they are and expand their horizons.

In domains like software coding, platforms like MIT's Scratch teach children to code through interactive, gamelike experiences. Scratch personalizes coding challenges and projects so that students develop computational thinking skills in a fun and engaging way. This early exposure to coding and problem-solving prepares students for careers in technology and other STEM fields.[11]

Integrating autonomous AI agents into primary and secondary education marks a new era in learning. By personalizing instruction, real-time feedback, and teacher support, AI can transform how children learn and prepare them for life's challenges. To realize the full benefits of AI in education, students, teachers, school boards, and parents must come together to address ethical and practical challenges.

Foremost, society must guarantee equitable access to AI technology. Second, AI developers must guarantee the impartiality, fairness, and accuracy of their tools. Transparency in modeling and training is critical. Third, school boards must train teachers to integrate AI effectively and meaningfully into classrooms, so that AI enhances pedagogy and student outcomes. Fourth, platform providers must protect student data from breaches and ensure their custody of it is transparent and ethical. Finally, society must properly value teacher-student relationships, which educators agree is critical to a well-rounded educational experience, and pay teachers more.

In a dystopian future, the less privileged may find themselves receiving basic AI-tutoring, designed for efficiency rather than depth, with oversight by systems rather than people. Meanwhile, the more affluent will enjoy even richer, bespoke education, their tutors still tailoring lessons to their individual learning goals with AI serving as sophisticated assistants. To guard against this future, we must be vigilant, so that innovators use AI to elevate the teaching profession, engage all students meaningfully, and address the needs of the most vulnerable. Only then can identic AI catalyze the improvement of education for everyone, everywhere, and foster a learning environment that respects human intelligence in all its forms.

## TRANSFORMING ON-THE-JOB LEARNING AND CAREER DEVELOPMENT

As industries evolve, so, too, must the skills and knowledge of their workforce. Manual laborers in the manufacturing sector need training in robotics, machine operation, and an understanding of AI-driven production systems. Healthcare professionals must stay current with the latest advances in health informatics, telemedicine, AI-based diagnostic tools, and data management. Transportation and logistics workers must hone their skills in managing and operating autonomous, data-driven logistics systems. In the retail sector, AI-powered chatbots and automation will reduce the need for traditional retail and customer service roles while increasing demand for digital and analytical skills.

The training and reskilling needs of the economy are genuinely huge. According to the World Economic Forum, over 50 percent of all employees need reskilling today because of the rapid adoption of new technologies.[12] Moreover, a report by McKinsey & Company highlights that 87 percent of executives are already experiencing skill gaps within their workforce or expect them within a few years.[13] Companies that invest in reskilling their employees are more likely to implement the latest tech solutions, more capable of adapting to changing marketplace dynamics, and less likely to lose ground or talent to competitors.[14]

Identic AI will usher in a new era of adaptive learning and transform how businesses train and develop their workforce. These intelligent systems promise to make learning more personalized, efficient, and accessible, reshaping how we grow and evolve professionally. Innovative platforms such as Docebo highlight the potential for AI companions to personalize learning pathways and training programs.

In 2005 in Milan, Italy, Claudio Erba founded Docebo with a simple yet ambitious goal: to make learning more accessible, engaging, and effective. "I saw that learning in the corporate world was often boring and ineffective," says Erba, reflecting on the early days of Docebo. "We wanted to create something that could adapt to each learner, making the process more engaging and more in line with how people learn."[15] What began as a small e-learning platform has since evolved into a global leader in learning management systems (LMS). Originally an open-source project, Docebo quickly gained traction and caught the attention of major corporations looking to modernize their training programs. By 2012, Docebo had transitioned to an SaaS-based LMS, which catapulted the company into the global market. Today, Docebo serves over two thousand companies in ninety countries, including giants like Walmart, Amazon, and Heineken.

At the heart of Docebo's success is an adaptive learning experience that harnesses artificial intelligence. Unlike traditional LMS platforms that offer a one-size-fits-all approach, Docebo's AI-driven system personalizes the learning journey for each employee. The platform analyzes data on how users interact with content, such as the types of resources they engage with, their performance assessments, and their feedback on the training material. With this data, the AI can tailor the pace, difficulty, and content of training programs to optimize outcomes.

As Erba explains, "Our AI engine continuously learns from user-generated data. It can predict what content a learner will need next and adjust the learning experience in real time. This level of personalization ensures that training is not only more effective but also more engaging."[16]

The platform also incorporates social learning elements, where employees can share knowledge, ask questions, and collaborate with peers. "Learning is most effective when it's collaborative," says Erba. "We've designed Docebo to support that collaboration, whether it's through forums, discussion boards, or peer-to-peer learning modules."[17]

The impact on corporate training is profound. According to Deloitte, companies using AI-driven adaptive learning platforms see a 30 percent increase in employee engagement and a 25 percent improvement in learning retention.[18] To move beyond fear of job displacements, companies must prioritize using AI to reshape work rather than replace workers. Deloitte claims leaders can address employee concerns by "carefully planning, investing in reskilling programs, and creating new opportunities that harness the unique strengths of humans and AI."[19]

## Real-Time Talent Intelligence and Skill Assessment

In a world where the workforce demands are constantly evolving, companies need real-time intelligence about the skills and capabilities at their disposal. Employees would also benefit from real-time feedback on their performance, allowing learners to understand mistakes, identify skill gaps, and improve quickly.

Eightfold.ai, a real-time talent intelligence company, has emerged as a game-changer in how companies identify, develop, and retain talent. Founded in 2016 by Ashutosh Garg and Varun Kacholia, both former Google and Facebook engineers, Eightfold.ai set out with a bold mission: "leverage the power of AI to provide the right career for everyone in the world."

Eightfold.ai began with Garg's deep interest in artificial intelligence and machine learning and his observation of the inefficiencies in the traditional talent management process. "I realized that there was so much untapped potential within organizations and that AI could be the key to unlocking it," says Garg. "Our goal was to build a platform that could match people to the right opportunities, not just based on their current skills, but on their potential and aspirations."[20]

This vision led to the creation of Eightfold.ai's talent intelligence platform, a sophisticated AI-driven solution that analyzes vast amounts of data to help organizations make smarter talent decisions. The platform provides real-time insights that go far beyond what traditional HR tools can offer by evaluating a candidate's or employee's entire career trajectory, including skills, experiences, and potential.

Eightfold's customers benefit from a more sophisticated approach to talent management thanks to the platform's unified view of the skills and capabilities across the organization. Companies can identify hidden talent within their workforce, make data-driven decisions about hiring and promotions, and redeploy talent internally when their business is re-organizing. If the talent intelligence analysis identifies skill or capability gaps, the platform automatically sources the required talent from a pool of high-quality temps, vendors, and contract workers.

For employees, Eightfold's distinguishing features include its ability to deliver real-time coaching and upskilling opportunities. The AI engine continuously analyzes employee performance, skill gaps, and career aspirations to recommend personalized learning paths and development opportunities. Ashutosh Garg explains, "Our platform looks at millions of data points to understand an individual's capabilities and potential. It then provides tailored coaching and upskilling recommendations that help employees grow in their careers while meeting the strategic needs of the organization."[21]

Suppose the platform identifies an employee on track for a leadership role. It might recommend specific training programs, mentorship opportunities, or project experiences to build the necessary skills. "The future of work is about adaptability," says Garg, who calls finding the right talent in today's competitive markets a trillion-dollar problem. "We are here to solve the trillion-dollar problem for the companies who spend millions every year just on hiring, to make sure that they not only hire quickly but hire the right talent and retain their employees as well."[22]

## Learning by Doing with Extended Reality

Traditional training methods have often fallen short in industries where hands-on experience is crucial, leaving a gap between theoretical knowledge and practical application. That gap is now closing, thanks to AI-powered simulations and immersive learning experiences that mimic real-world scenarios. With extended reality (XR) technologies, learners could practice skills in a safe and controlled environment and apply their knowledge in practical situations.

Running predictive analytics on market trends to anticipate needed skills, an AI companion could simulate real-time scenarios or guide users by their hands to develop what the workplace needs most. For example, in an AR-enabled workspace, the AI companion could lay step-by-step instructions over the user's physical environment so that, say, an engineer could assemble the machine in front of her. The AR interface could highlight parts, show assembly techniques, and flag mistakes in real time. Such feedback in context can improve accuracy and speed in learning new tasks.

Alternatively, an AI companion could immerse learners in VR simulations where they can practice skills in lifelike scenarios. For instance, a medical professional could perform virtual surgeries or emergency procedures in a risk-free environment, with the AI companion preempting errors and suggesting improvements. This kind of immersive training accelerates learning because learners experience and interact with the environment as if it were real.

Does your organization need to strengthen its collaborative capacity? AI companions could facilitate team-based scenarios so that team members cultivate those skills in a controlled virtual space. The AI companion could monitor interactions, share observations of team dynamics, and suggest ways to improve communication or resolve conflicts, particularly in leadership and management training.

We see glimpses of these in platforms like Strivr, a pioneer in leveraging XR technologies to transform how industries like manufacturing, retail, healthcare, and sports prepare their workforce for the real world. Founded in 2015 by former Stanford University football coach Derek Belch, Strivr started as a VR tool for training athletes and improving player performance. "As a coach, I saw firsthand how much players struggled to translate what they learned in the classroom to the field," Belch recalls. "I wanted to find a way to give them more realistic practice without the physical toll."[23]

Belch teamed up with Stanford's VR lab. Together, they created simulations for athletes by combining VR, AR, and mixed reality (MR). So successful were these early experiments that they decided to apply the technology far beyond sports. Belch launched Strivr to expand his sports training tool into a platform for enterprise-level training in industries where traditional training methods were inadequate and where safety, precision, and hands-on skills were critical.

Today Strivr delivers scenario-based training that replicates what employees face in their jobs. "Traditional training often relies on written manuals or videos, which can't fully capture the nuances of real-world tasks," says Belch. "With Strivr, we can immerse employees in the exact situations they'll encounter on the job, helping them build muscle memory and confidence in a way that no other training method can."[24]

Strivr's training modules are highly interactive: Employees must complete cognitive and physical tasks. The AI technology tracks their movements and gives real-time feedback and analytics so that organizations can identify skill gaps and plan subsequent trainings. This learning-by-doing approach improves retention and reduces the risk of costly mistakes in high-stakes environments. For example, the global retailer Walmart has adopted the platform to train employees in customer service, emergency response, and new technology rollouts. By immersing employees in realistic scenarios, Strivr has helped Walmart reduce training times and improve performance across its vast network of stores. In the sports world, NFL teams use Strivr to prepare players for games; athletes can visualize plays and practice reactions without physical wear and tear.

Strivr's approach shifts how industries approach skill development. By leveraging XR technologies, organizations can onboard new employees and develop existing staff more effectively, efficiently, and collaboratively than ever before. According to PwC, employees trained with VR-based tools learned up to four times faster than those trained in classrooms.[25] Wow. AI and XR technologies can do more than enhance what and how employees learn. As these capabilities mature, training and development innovators will transform the world of training.

## The AI Mentor and Your Personalized Career Path

In today's job market, career development is daunting. The sheer number of options and the rapid changes in salable skills can overwhelm job seekers across industries. But platforms like Docebo, Eightfold, and Strivr currently serve organizations, not individuals. So how could our personal autonomous AI agents help shape our career trajectory over our lifetime?

Think of this AI companion as a digital mentor who understands your skills and interests and tailors advice to your career aspirations. Such AI-driven career guidance could transform how professionals plan and navigate their career journeys and align their goals with the demands of the modern workforce.

This AI companion would draw on a lifetime of data, including a person's work history, education, achievements, and even personality traits, and holistically analyze that individual's skills, interests, and career aspirations to work up a career profile. Then it could personalize a career pathway with a step-by-step guide to transitioning into a desired role or advancing in a current field. These pathways would continuously update based on the individual's progress, market trends, and emerging opportunities.

Suppose a software engineer wants to pivot into artificial intelligence; the AI companion might suggest a learning path with specific courses in machine learning, hands-on projects in AI, and networking opportunities with industry professionals. The AI might also highlight lateral moves within the individual's current organization to gain valuable experience and connections while maintaining career stability. Finally, the AI could predict high-demand skills, emergent roles, and the most promising career trajectories.

If the AI detected a rising demand for cybersecurity experts, then it might advise a network administrator to pursue certifications in cybersecurity and opportunities in that field. The AI can also offer insights into salary trends, job stability, and growth potential in various roles, helping professionals weigh the pros and cons of different career paths.

Having identified a career path, the AI could identify job openings, freelance projects, or internal promotions that match the individual's skills and aspirations. It could even tailor resumes and cover letters to highlight relevant experiences and achievements, increasing the chances of landing desired roles.

With the AI companion's data-driven pathway, professionals could make informed decisions based on real-time analysis of industry trends, job market demands, and personal strengths that put their needs and interests first, not necessarily those of current or prospective employers.

## EXPLORING NEW FRONTIERS: LIFELONG PERSONAL GROWTH

Imagine embarking on a journey of personal growth with an unwavering companion by your side who is always available, infinitely patient, and incredibly knowledgeable. This vision of a fully bespoke personal growth coach is the promise of AI companions: a complete transformation of how we learn new hobbies, acquire languages, pursue fitness goals, and explore ideas. In this new era, personal growth is becoming a collaborative venture with intelligent machines that understand our needs, preferences, and aspirations and help us achieve our goals more effectively and meaningfully. Your digital self will likely include a creativity coach that helps you brainstorm and refine ideas, a mindfulness guide that recognizes stress patterns and recommended calming exercises, and a networking adviser that prepares you for social interactions and expands your professional reach. By augmenting our digital selves with these AI-driven capabilities, we're not just pursuing self-improvement; we're cocreating our future selves with an intelligent partner.

## Mastering New Hobbies

Thanks to television's gurus of how-to—Julia Child, Bob Ross, Martha Stewart, and Bob Vila, to name a few—we have grown up with virtual experts in our kitchens, garages, and craft studios. Today AI companions like Yousician for music or Tasty for cooking personalize their guidance and adapt their lessons to each hobbyist's pace and progress. They offer instant feedback, steer budding chefs clear of culinary no-nos, and suggest new techniques. Such personalized instruction turns even the most challenging hobbies into rewarding pursuits.

Let's start with Yousician, an interactive music education platform for learning instruments like guitar, piano, ukulele, bass, and voice. Founded in 2010 by Chris Thür and Mikko Kaipainen in Finland, Yousician combines AI and gamification in a learning experience that adapts to each user's pace and skill level.[26] In essence, Yousician listens to users play through their device's microphone and offers instant feedback, as if a human music teacher were right there, nodding along.

Thür and Kaipainen envisioned a world where anyone could pick up an instrument and learn to play without such barriers as cost, time, or fear. Their app is intuitive and fun, with rewards, challenges, and progress tracking. Mastering music feels more like a game, one that millions of music makers are playing worldwide. As co-founder and CEO Chris Thür explains, "We want to make learning how to play the guitar, the piano, or even singing, as democratic, accessible, and easy as possible, and we want to bring our mission into every household around the globe."[27]

## Acquiring New Languages

Research has shown that learning multiple languages from a young age can enhance cognitive abilities, such as problem-solving, multitasking, and memory.[28] The average European speaks two languages fluently, and many Europeans are fluent in more than two.[29] This rate of multilingualism is higher than the average in North America, where most people, particularly those born in the United States, speak only one language—English. The average is slightly higher in Canada, where there are two official languages, English and French.

In languages, Europeans have a clear advantage with many different tongues in close proximity. Many Europeans hear multiple languages from a young age, especially in border regions.

What if AI companions exposed everyone to multiple languages from birth, simulating the environment of children growing up in multilingual households? This continuous exposure would likely make bilingualism or multilingualism the norm rather than the exception, with its cognitive benefits more widespread.

Numerous AI-powered language apps already offer personalized language services. For example, Duolingo and Babbel use AI to track each learner's progress, identify areas of difficulty, and adjust the curriculum accordingly. Imagine conversing in Spanish with an AI companion, which interjects real-time corrections and reinforces learning, so that learners acquire new languages faster and more intuitively.

## Pursuing Personal Fitness

Getting in shape and staying fit require discipline, motivation, and a personal plan. Today, many people use fitness apps, join sports clubs, hire personal trainers, follow social media influencers, participate in online forums, and consult health websites. Many also rely on wearable devices that track and analyze their physical activity. However, the quality and accuracy of the advice varies greatly by source.

AI companions have the potential to transform how people receive and act on fitness advice, making it more personalized, accessible, and effective. AI companions can analyze an individual's unique data, including health history, dietary habits, fitness level, goals, and preferences, to create tailored workout and nutrition plans. Unlike generic advice, AI can adjust these plans in real time based on the user's progress, performance, and feedback, ensuring that the advice is always relevant and practical.

Motivation is a significant factor in maintaining a fitness routine, and AI companions could play a crucial role here. By providing real-time encouragement, reminders, and personalized motivational messages, AI companions can help users stay on track. They can also learn what motivates the user best, be it gentle nudging, competitive challenges, or rewards, and adapt their approach accordingly.

When paired with wearable fitness tracking devices, AI companions could provide real-time feedback during workouts, correct form and posture, and detect stress points before injuries occur. With AI, wearables could do more than track; they could act as virtual personal trainers.

Often, the best part of exercise is the social component that comes with group workouts, like running, cycling, or hitting the gym with friends. AI companions could facilitate connections with others with similar fitness goals, creating virtual communities where users can share progress, challenges, and successes. AI companions could also democratize access to high-quality fitness advice. While personal trainers and expert consultations can be expensive and inaccessible to many, AI companions can offer expert-level guidance at a fraction of the cost, making personalized fitness coaching available to a broader audience.

# LIFELONG LEARNING IN THE AGE OF AI: OPPORTUNITIES, CHALLENGES, AND THE ROAD AHEAD

Ubiquitous use of identic AI in education raises profound questions. Will AI-enhanced learning systems exacerbate existing inequalities by favoring those with better access to technology? Can we trust AI-driven systems to remain free of bias, ensuring that personalized learning paths do not unintentionally reinforce social or economic divides? Furthermore, as education becomes more reliant on AI companions, how will human educators balance the use of these technologies with the need to foster creativity, critical thinking, and emotional intelligence?

It's worth dwelling on this last question. One of the more salient critiques of AI in education revolves around the concern that AI could erode students' ability to learn and retain information. If students can rely on AI for instant answers, they may bypass the cognitive processes necessary for deep learning, such as critical thinking, problem-solving, and long-term memory formation. Just as reliance on calculators initially sparked worries about students losing basic arithmetic skills, the presence of AI in the classroom raises questions about how it might impact foundational knowledge and intellectual discipline. If misused, AI could encourage passive consumption of information rather than active engagement with it, creating a generation of learners who lack the ability to process, analyze, and internalize knowledge independently.

However, this critique underestimates the potential of AI to complement and enhance human learning when used thoughtfully. AI doesn't have to be a crutch; it can serve as a tool to deepen understanding and foster intellectual curiosity. The key lies in designing AI systems that prioritize learning processes over immediate answers, ensuring that students remain active participants in their education.

Ultimately, parents, educators, business leaders, and technologists must collaborate within the learning ecosystem so that AI augments human connection, creativity, and empathy in education. The question is not whether AI will transform learning but how we will shape this transformation to benefit society as a whole.

Chapter 6

# CREATIVE INTELLIGENCE: IDENTIC AI IN ART, MUSIC, AND MEDIA

For nearly three decades, digital media and the internet have disrupted the traditional business model for arts, music, and culture. Once the cornerstone of revenue, physical sales have been eclipsed by downloads and streaming services for music, books, and art. Streaming is a give-and-take business: It gives fans unprecedented accessibility and takes away artists' per-unit earnings.

Pirated or user-generated content platforms further eroded the intellectual property (IP) royalties that artists relied on to pay their bills and make their art. Algorithms trained on social media data and unleashed on social media users reshaped their behavior as consumers. Like fast-food franchises of the mind, the sites shoveled bite-sized and ad-wrapped chunks of content into users' feed bags. Tastemakers like publishers, record labels, and galleries lost influence as creators turned to self-publishing and crowdfunding, disintermediating the content food chain and democratizing access. Everybody showed up. Algorithms helped choose the winners.

Ultimately, the digital revolution fractured the elite ecosystem of the arts. Veteran creators had to navigate another complex terrain, one foreign to their agents, publicists, and business managers. New creators seized opportunities to reach global audiences, experiment with direct-to-fan models, and harness AI and blockchain to manage their rights and revenues. In this chapter, we show how Web3 supports a new business model for culture.

At the same time, the growing role of AI in creative industries raises concerns. Could relying on an AI companion for technical expertise and inspiration erode your skills and dull your mind, turning you into a passive collaborator? Will your AI-generated content lack human authenticity and emotional depth? Perhaps the convenience of AI companions will stifle spontaneity and intuition, and algorithms rather than inspiration will drive creative processes.

The growing use of AI in creative industries also brings legal and ethical challenges. For example, who—if anyone—owns AI-generated work? And will society come to prefer AI-driven creativity to human originality? As we dive into AI-enhanced creativity, we fear a scenario where the rising tides of machine-generated output overwhelm the scattered atolls of unique human voices.

Of course, we make no idle speculations. Identic AI has begun reshaping the creative arts, music, and news reporting. These advances promise to disrupt how we create, consume, and perceive cultural content. Must we redefine creativity, authorship, and authenticity? Or will AI redefine them for us?

In the realm of visual arts, AI made waves when the first-ever AI-generated portrait sold at auction, *Portrait of Edmond de Belamy*, fetched a whopping $432,000 at Christies in New York.[1] Programs like DALL-E and Synthesia can generate images in the style of Picasso or create short films from textual descriptions. Which begs the question: What is the nature of creativity? Are developers and trainers of AI models the truly creative ones now? Or is AI the paintbrush, the palette, and the canvas combined, where creative users of AI tools are the real artists? Can a machine truly create, or does it merely mimic the masters? When an AI-generated piece garners acclaim, does the machine, its creators, its source material, or the algorithm deserve the credit? The provocative emergence of AI in the art world forces us to reconsider the boundaries of artistic expression and the essence of human creativity.

Music, too, is undergoing a transformation. OpenAI's MuseNet and AIVA (artificial intelligence virtual artist) compose music that spans genres and styles, from classical symphonies to contemporary pop hits. One could easily tailor these AI compositions to evoke specific emotions or fit particular moods, offering unprecedented personalization in music creation. Imagine your AI companion modeling the music libraries of your favorite bands to spin up new tracks on the fly. AI music engines could routinely produce the next chart-topping singles. In fact, Max Richter, the groundbreaking contemporary composer, claims that AI-generated music is already climbing the charts, but no one is openly admitting it.[2] This revolution in creating music could democratize access to high-quality composition tools. At the same time, it could challenge the traditional roles of musicians and composers.

With access to vast amounts of data and the capacity to generate coherent, fact-based reports, AI-powered systems like OpenAI's GPT-4 may also transform journalism. Digital journalists could produce accurate articles on complex topics in a fraction of the time that human counterparts take. But would its "investigations" and "analysis" be as rigorous and ethical as human journalists' efforts? Would AI care about the implications of its public revelations? Could the subjects of its "exposés" sue it or its steward for libel? How do we verify the accuracy of algorithmically generated news? As newsrooms use AI more, they must balance efficiency and integrity.

Integrating AI into the creative arts, digital entertainment, music, and journalism heralds change and challenges in the training and work of professional journalists, artists, musicians, and other creators. The key will be for creators to harness AI's capabilities while preserving the human elements that enrich and define our cultural experiences. The story of creative AI is not just about technological advancement; it is about re-imagining and perhaps revaluing human expression amid the creativity of intelligent machines.

## HOW AI COMPANIONS ARE REDEFINING ART, FILM, AND BEYOND

For decades, skeptics have insisted that technology will never truly originate or create, computers will never learn autonomously, and robots will never surpass human beings in repetitive manual labor. Yet, in the past ten years, groundbreaking advances in artificial intelligence have steamrolled the naysaying.

Take, for instance, tools like Midjourney and Stable Diffusion. These AI programs can generate astonishingly realistic portraits of your children in the styles of masters like Rembrandt or Van Gogh. With these AI tools, anyone with a computer can make art. Amateurs and curious consumers can experiment with artistic styles and create art that sates their tastes and values. For professionals, the AI tool set opens creative avenues without roadblocks. Let's explore the road ahead for artists, filmmakers, and audiences.

### Artistic Collaboration and Technological Exploration

In the digital art landscape, Mario Klingemann has broken new ground at the intersection of artificial intelligence and creative expression with generative adversarial networks (GANs).[3] According to Klingemann, he combines "a coder's analytic mind, an artist's creative fervor, and a dash of mad scientist."[4] In his studio, Klingemann prefers neural networks, code, and algorithms to sable-hair paintbrushes.

Technology's potential to expand human creativity has long fascinated

Klingemann, born in Germany in 1970. Early in his career, he explored digital media, but the advent of machine learning technologies, particularly GANs, brought him new tools for his artistic pursuits. With GANs, a class of AI algorithms that can generate new data from existing datasets, Klingemann could create innovative as well as thought-provoking art.

In 2018, Klingemann's GAN-based installation *Memories of Passersby I* opened at the Espacio SOLO in Madrid.[5] The work consists of two monitors and a computer running a Klingemann-trained algorithm that generates an endless stream of human faces, each distinct and ephemeral.[6] The work exemplifies his approach: He used AI software as a tool to train an algorithm as his apprentice in the creative process. The installation challenges viewers to consider the impermanence and subjectivity of identity. *Memories of Passersby I* was also among the first AI artworks that art connoisseurs bought (for £40,000, approx. $52,635 at the time) in a traditional auction house, Sothebys.[7]

Klingemann's work often explores where human creativity ends and machine creativity begins. He trained these systems to produce new works that can both surprise and unsettle. But can a machine genuinely create, or does it mimic the creativity of its human creators? Or does it monitor and learn from critics' and consumers' responses to its art as they pass by?

In one of his projects, *Neural Glitch*, Klingemann fed neural networks with images and videos to create abstract, often distorted outputs.[8] These works play with "glitch art," which celebrates rather than corrects errors and distortions. In Klingemann's hands, errors and randomness are integral to the creative process.

The art world has recognized Klingemann's contributions to the field through exhibitions at the Centre Pompidou in Paris, the Photographers' Gallery in London, and the Ars Electronica Festival in Austria. In 2018, he received the Lumen Prize as a leading figure in AI-mediated art.[9]

Mario Klingemann continues to push the artistic capabilities of AI. He is exploring more complex forms of human-machine collaboration, where AI not only generates but also curates and critiques art. Such advances could lead to new forms of interactive installations where the audience's reactions influence the artwork in real time.

Klingemann reminds us that the intersection of AI and art is not just about new tools but about the creative process itself. In this sense, he has much in common with Sougwen Chung, a Chinese-Canadian artist pivotal in the fusion of art and technology. Like Klingemann's, her work frolics along the traditional boundaries between organic and plastic, human and nonhuman, physical and virtual.

Chung began using AI in the early 2010s. Initially trained in fine arts, she transitioned to digital media so that she could explore the symbiosis between human and machine, a theme now central to her work.

In 2015, Chung debuted *Drawing Operations*, a hallmark of her practice. Chung trained a robotic arm with machine learning algorithms to collaborate with her in creating intricate drawings. The robot named D.O.U.G. learns from Chung's work, upgrading its style and techniques through that interplay.[10]

Over time, the robot's algorithms develop its distinct Chung-influenced style. Her more recent works feature new human-machine configurations that connect to biofeedback. As she paints, Chung wears an EEG headset to track her brain waves, and her robotic assistants paint the undulations of her brain.[11] Chung is both master and muse.

## AI Companions and the Digital Brush

Chung and Klingemann evoke questions about authorship, creativity, and the artist's role in shaping how society views and values their work. The two artists showed us how AI companions can augment and even mimic the creative process. Advanced algorithms and machine learning systems can analyze vast amounts of visual and auditory data and replicate or mash up the styles and techniques of different artists to produce derivative works.

Even nonartists can participate. DALL-E, an AI developed by OpenAI, can generate images from the text prompts of people who ask for blends of concepts, styles, and objects. The more imaginative the people, the more imaginative the prompt and the result. Likewise, users can text prompt Midjourney, another AI platform, to create visual art. Rather than paint by number, users prompt by words, and these tools do the rest. We all can produce art again, perhaps reminiscent of those primary school days when we rocked our inner Picasso or, more likely, our Dr. Seuss. In other words, just about anyone can wield a digital brush.

Some fear an oversaturation of the art market, where the volume of created works diminishes the perceived value of individual pieces.[12] If anyone can produce high-quality images with minimal effort, then why do we need so-called trained professional artists?

Of course, identic AI may open up new markets for art. Digital platforms can curate AI-generated artworks tailored to individual tastes. Art can span the spectrum of rarefied object to drive-through commodity. In the end, Chung and Klingemann are still integral to their art and its value. Their choices and their stories still interest and engage people.

# THE SOUND OF TOMORROW: HOW IDENTIC AI IS TRANSFORMING MUSIC

On the music scene, few artists are as forward-thinking as Grimes. Known for her ethereal vocals and experimental electronic beats, Grimes has added artificial intelligence to her creative process. Her innovative approach represents a significant shift in how musicians create in the digital age.

## How Grimes Is Pioneering New Soundscapes

Grimes, whose real name is Claire Boucher, has experimented with various AI technologies to create new music. One of her primary tools is the AI platform Elf.Tech, a real-time collaborative music creation platform that incorporates algorithms to generate musical suggestions.[13] On this platform, Grimes can explore, sample, or adapt a virtually limitless array of sounds, rhythms, and melodies.

First, Grimes inputs her musical themes and preferences into the AI system. Next, the AI applies algorithms to her work and identifies the patterns and motifs that define her unique sound. Then, it generates compositions with chord progressions, beats, and melodic lines in line with her artistic vision.

With AI, she can discover ideas that might not have occurred to her otherwise and experiment with unusual time signatures, harmonic structures, and sound textures. She can ask AI to draw from classical music to contemporary genres so that she has a rich portfolio of sounds to work with.

In addition to generating music, the AI assists with composing lyrics. By analyzing keywords and themes that Grimes supplies, the AI can suggest lyrics that complement the music. Grimes can explore those lyrical phrases, concepts, and themes in new ways and think outside her own patterns.

AI tools also extend to visuals. Grimes has incorporated AI-generated visuals in her music videos and live performances. Generated by AI in real time, the visuals react to her music, immersing audiences in dynamic experiences that heighten the artist's presentation.

Grimes extended her AI experiments to her fan base by granting them permission to use AI-generated versions of her voice in creating new music. Through her partnership with Elf.Tech, her fans can synthesize new tracks featuring her voice without complex recording setups.[14] With this project, Grimes sought to expand her artistic reach and empower others to explore their creativity. She has pioneered a new form of collaboration where communities drive the music making. In exchange, Grimes receives a 50 percent royalty on any AI-generated song recording that uses her voice.[15]

Grimes's use of AI in her music is also a philosophical exploration of the relationship between human creativity and machine intelligence. She views AI as a partner, one that can challenge her artistic instincts in the creative process. By integrating AI into her workflow, Grimes pushes out the boundaries of her music and sets a precedent for other artists.

However, using AI to write lyrics and music raises important questions about the nature of creativity and authorship. While AI can generate musical ideas, the human artist must decide whether and what to do with them. To Grimes, AI is a tool—albeit a sophisticated one—that assists in the creative process rather than replaces human creativity.

The more Grimes innovates with AI, the more she helps scope out a new frontier in the music industry. From composition and recording to distribution and consumption, AI is redefining roles and disrupting the value chain. For fans and fellow musicians, Grimes lets us sample the future of music.

## Your Personal Music Connoisseur

With AI, music consumers can personalize their listening experiences. Today's streaming platforms use algorithms to analyze user preferences and habits and curate playlists for individual tastes. Consider Spotify's Discover Weekly playlist. Each week, the platform generates a unique playlist for every user based on the user's listening history and the preferences of users with similar tastes. This feature has become immensely popular. Listeners enjoy discovering new music and artists.

Future AI services could dive deeper into niche genres and artists, expanding listeners' repertoires. Call it your personal music connoisseur. The music discovery algorithms could benefit emerging artists or genres as users explore new sounds. AI systems could analyze listening habits, lyrics, instrumentation, and production styles to find nonobvious matches. In doing so, AI systems could harness trends and patterns to predict which songs and artists will likely become popular before their release.

As our AI companions evolve, they could analyze a broader set of data, including mood, context, time of day, and physiological data like heart rate or stress level (if the user consents). With this analysis, streaming services could highly personalize playlists to align with a user's emotional state or activity, such as working out, studying, or relaxing. Your morning playlist might feature upbeat tracks to energize you, while your evening playlist might run calming tunes to wind you down.

Integrating your streaming service with calendar apps, fitness trackers, weather reports, and smart home devices could match your playlist to your routine, activity, location, and other invitees. If you skip a song or prefer specific genres or tempos, the algorithm can adjust the playlist. Blending AR/VR with music could lead to immersive listening experiences. Maybe the next time you fire up Jimi Hendrix, you could relax on the couch while attending his virtual concert.

## The Future Soundscape

The future of music promises highly personalized, dynamic, and immersive experiences where we discover, interact with, and enjoy music, making it a more integral and responsive part of our daily lives. But let's consider in more detail what this means for musicians. Artists like Grimes are showing the world how AI can augment and sometimes replicate the creative process. But what if an AI composer like AIVA learns the intricacies of composers like Beethoven and Mozart and originates compositions that echo their styles? When advanced algorithms and machine learning systems can analyze such vast quantities of musical data (basically, the total of recorded music), it's not inconceivable that many will consider AI compositions superior.

Plenty of high-profile musicians and artists have rejected this vision. In April 2024, over two hundred artists, including Katy Perry, Billie Eilish, Pearl Jam, Nicki Minaj, and Elvis Costello, called irresponsible AI practices "an assault on human creativity" and signed an open letter urging tech companies not to sabotage hardworking musicians.[16] The signatories acknowledged the potential for AI to advance human creativity. But they strongly objected to the rampant use of AI models trained on unlicensed music, which they called "efforts directly aimed at replacing the work of human artists with massive quantities of AI-created 'sounds' and 'images' that substantially dilute the royalty pools ... paid out to artists."[17] The signatories warn that current practices could ruin "many working musicians, artists, and songwriters ... trying to make ends meet."

UK-based singer-songwriter Nick Cave called ChatGPT's lyrics "a grotesque mockery of what it is to be human."[18] Cave penned the lament on his personal blog after a fan sent him ChatGPT-generated lyrics "written in the style of Nick Cave." As Cave put it, "Songs arise out of suffering, by which I mean they are predicated upon the complex, internal human struggle of creation and, well, as far as I know, algorithms don't feel."[19] Indeed, the AI-composed lyrics were no match for Cave's original works.

On the other hand, musicians like Daniel Bedingfield have argued that AI is music's future, and anyone who fights it faces obsolescence. "AI is now here forever," he told *The Guardian*. "And so I think that there will be two paths: there'll be the neo-luddite path, and then there'll be everyone else, most of the planet, who thinks the music's really good and enjoys it."[20] Bedingfield doesn't see AI replacing human musicians. Instead, he uses AI as a collaborative tool to explore new sounds, experiment with different styles, and break through writer's blocks. Using platforms like Udio and OpenAI's MuseNet, musicians like Bedingfield can input a melody or chord progression, and the AI will generate an entire arrangement, providing inspiration and new directions for their work. By integrating AI into their creative processes, artists can grow and explore all-new musical territories. "Why fight AI," asks Bedingfield, "when you can have a whole gospel choir singing your chorus in two days' time?"[21]

AI is also transforming the production side of music. AI-powered tools can assist in mixing, mastering, and even lyric writing, streamlining the production process and ensuring high-quality results. LANDR's AI-driven mastering service analyzes tracks, equalizes levels of frequencies of sounds, and compresses the range of volumes to polish the work. LANDR may not sideline the world's best human producers. However, it could help independent artists and smaller studios to make professional-quality demo tracks for select release.

The most likely scenario is a hybrid future in which AI companions and human musicians collaborate to create new and exciting music. AI can handle the technical and repetitive aspects of music creation; human artists can focus on the creative and emotional elements.

## THE NEWSROOM: HOW IDENTIC AI IS TRANSFORMING NEWS MEDIA

Tax evasion and offshore banking have long been global challenges with far-reaching implications. Through these practices, individuals and corporations had hidden vast sums of money, depriving governments of critical revenue for public services like healthcare, education, and infrastructure.[22] The use of offshore accounts and shell companies creates a shadow financial system that operates beyond the reach of standard regulatory oversight. These schemes not only exacerbate economic inequality but undermine the social contract because ordinary citizens often bear heavier tax burdens while the wealthy evade their fair share.[23]

In 2016, a "John Doe" leaked to the press what would become known as the "Panama Papers," documents from the Panamanian law firm Mossack Fonseca that exposed how politicians, business leaders, and celebrities use offshore banking to conceal wealth and avoid taxes.[24] On one hand, the leak highlighted how these business practices contributed to fraud, money laundering, and financing of other illicit activities worldwide. These revelations underscored the urgency for transparency and accountability in the global financial sector.

At the same time, the Panama Papers investigation spotlighted the use of artificial intelligence in investigative journalism. The leak of 11.5

million documents and 2.6 terabytes of data from Mossack Fonseca posed an unprecedented challenge for journalists.[25] First, the data was complex: Mossack Fonseca created and administered an intricate web of 214,488 offshore structures between 1977 and 2015. Second, the data was voluminous: It contained information ranging from internal and external emails to databases and millions of scanned documents. To sift through this enormous digital pile, journalists applied advanced technologies, including AI.

Initially, the leaked data landed with journalists at *Süddeutsche Zeitung*, one of Germany's largest daily newspapers.[26] *Süddeutsche Zeitung* soon realized that the data was more than it could handle, so it turned to the International Consortium of Investigative Journalists (ICIJ) for help. In turn, the ICIJ enlisted 370 independent journalists from more than 100 media partners in over 80 countries. Even then, 370 intrepid journalists could never hope to fully extract the truth from a trove of 11.5 million documents on the opaque world of offshore finance.[27] So the ICIJ turned to a suite of AI tools and techniques to assist in data processing, analysis, and pattern recognition.

One of the primary tools was Nuix, an AI-powered data analytics platform for indexing and searching vast amounts of unstructured data. Using Nuix, journalists quickly located critical documents, emails, and financial records by searching for specific names, dates, and keywords.

Another critical AI tool was Linkurious, a data visualization platform for mapping and analyzing relationships among entities in documents. Linkurious helped visualize the connections among shell companies, their owners, and other parties. By tracing the flows of money through various offshore structures, they identified behavioral patterns and uncovered the ultimate beneficiaries of these tax avoidance schemes.

A third technology was natural language processing (NLP), which extracts meaningful information from vast volumes of text. Without NLP, journalists would have struggled to identify key players, grok the context of financial transactions, and decipher the legal and financial jargon that the parties used to obfuscate the true goals of their activities.

A fourth tool was the open source Apache Tika, a framework for detecting and extracting metadata from a thousand different file types, which journalists used to sort out the most significant aspects of the data, thereby streamlining their investigation.

Of course, AI technology has advanced significantly since the ICIJ completed its Panama Papers investigation. Since then, the ICIJ has been studying how to deploy machine learning algorithms in investigations. For example, by training algorithms on known instances of illicit financial activity, journalists could detect anomalies and unusual patterns in financial transactions, highlighting potential fraud, money laundering, and tax-evasion cases. Marina Walker Guevara, who has been leading this work for the ICIJ, raised such questions as:

> What would our research look like if we were to deploy machine learning algorithms on the Panama Papers? Can we teach computers to recognize money laundering? Can an algorithm differentiate a legitimate loan from a fake one designed to shuffle money among entities? Could facial recognition more easily pinpoint which of the thousands of passport copies in the trove belong to elected politicians or known criminals?[28]

According to Guevara, the answer to all those is yes. Even without the latest AI advances, the collaboration between investigative journalists and AI technologies was critical in the Panama Papers investigation. As ICIJ's executive director, Gerard Ryle, explained:

> This was the largest leak in history. It was also the largest journalism collaboration in history. We are dealing with a new phenomenon where whistleblowers can gather information on a scale never thought possible. Journalists have to turn to new technologies to understand what we are seeing and start looking for patterns because we find stories in patterns. We don't find them in individual names or individual pieces of information, and finding patterns is where the software comes in.[29]

Moreover, with those AI tools, the ICIJ and its partners deciphered the massive dataset in a fraction of the time they would have taken using traditional methods to uncover the extensive network of offshore entities and the individuals behind them. Their findings prompted widespread public and governmental scrutiny and numerous legal and regulatory actions worldwide. As a result, governments have recouped more than $1.2 billion in fines and back taxes.[30] Experts consider this figure an understatement due to disclosure limitations in many countries such as Poland and South Korea, which have not put a number on tax monies they have recovered.

## A New Era of News Creation

In the digital age, where information flows at lightning speed, journalism's role has never been more crucial—or more challenging. Identic AI is poised to revolutionize the creation and consumption of news media. Generative AI has already automated writing. With remarkable speed, platforms like OpenAI's GPT-4 can generate news stories, sports recaps, financial reports, and weather forecasts in response to user prompts.

AI capabilities are very apparent in financial news reporting. Before the adoption of AI, a large team of Associated Press (AP) journalists manually created quarterly earnings reports for around three hundred Standard and Poor's 500 companies. This labor-intensive and time-consuming process often led to delays in reporting corporate results. Today, AP and other news organizations use generative AI and data from Zacks Investment Research to generate concise earnings-related articles within minutes.

In theory, with such efficiency in earnings reporting, human reporters should have time for more in-depth and investigative journalism. But scores of unemployed journalists tell another story. One recent study estimated nearly ten thousand AI-related layoffs in the United Kingdom and North America in 2023 and 2024, likely an undercount.[31]

As we move from generative AI to identic AI, automated news agents will increasingly improve in sifting through reams of data in real time, identifying trends and generating insights. This capability has already transformed how news organizations handle breaking news and ongoing stories. For example, Reuters designed its Tracer AI tool to scan social media platforms like Twitter/X and other online sources to identify potential breaking news stories.[32] Tracer analyzes the data to gauge the credibility of sources, the relevance of information, and its potential impact on emerging stories.

As early as 2016, *The Washington Post* developed an AI-powered tool named Heliograf that automatically generates news stories.[33] Heliograf monitors real-time data feeds on social media, election results, and sports scores to generate and update news articles quickly. During the 2016 US elections, Heliograf monitored election results from government databases and updated news on key races in real time. In 2020, *The Post* added an AI voice assistant to insert audio election updates into its political podcasts automatically.[34]

Since 2020, countless new applications for auto-generated news have showcased AI's ability to process data quickly so that news outlets can respond to fast-moving events with unprecedented speed. During the chaotic early hours of the 2021 US Capitol riot, Reuters's Tracer sifted through a sea of social media posts and identified key developments and credible sources.[35] With Tracer, Reuters could deliver timely and accurate updates, keeping its audience informed as events unfolded.

Similarly, during the 2022 Russian invasion of Ukraine, news organizations used AI tools to monitor official statements, social media updates, and on-the-ground reports and to verify information quickly, track troop movements, and report on human rights violations, ultimately publishing comprehensive coverage of a highly dynamic and volatile situation.[36]

AI is also creating multimedia content. AI systems can generate audio, video, and interactive graphics for immersive news experiences. News organizations like *The New York Times* use AI to create video summaries of news stories.[37] These AI-generated videos can condense long articles into short clips so that audiences get the gist on the go.

During the Israel-Hamas conflict, *The New York Times* used AI to analyze satellite imagery of Israel's bombing campaign in southern Gaza. The AI tool scanned the satellite images for bomb craters and detected over 1,600 potential holes. A team of journalists then manually reviewed these images to identify those most likely created by 2,000-pound bombs, providing a comprehensive visual and data-driven account of the conflict.[38]

## Your Personalized News Companion

The innovations in professional news reporting highlight one side of the story. What if everyone had access to a powerful AI companion with the same analytical capabilities as the ICIJ to unravel, say, the Mar-a-Lago papers or Hunter Biden's laptop?

We already know that machine learning algorithms analyze user behavior to personalize content for readers. This personalization ensures that readers receive relevant news. But your AI companion will really get personal.

Let's say the fast-paced world of technology and innovation interests you. Your AI companion can analyze press releases, tech blogs, and patent filings to inform you of the latest AI developments, robotics, software, and company news. Sports fanatics will ask their personal AI to produce match summaries, player profiles, and pre- and post-game analyses. Or maybe you want your AI to analyze government documents, policy papers, and political speeches to identify patterns and summarize new laws, regulatory changes, and the latest political machinations.

AI systems have great capacity to distill long articles into bullet points so that we can grasp key points quickly. That's a chicken nugget for people who prefer to consume news in bite-sized formats. Or perhaps you really like to dig in. Your AI companion could analyze and understand the context of news stories, profile public and private sector leaders, and assess the risks of world events and their economic and political implications.

With multimodal AI, you can choose the medium and format of your news. Whether you prefer reading articles, watching videos, or listening to audio, these AI systems can serve up the news in your preferred mode of intake. Popular AI-powered voice assistants like Amazon's Alexa, Google Assistant, and Apple's Siri already tailor news briefings to users' preferences so that they can stay informed while driving, cooking, or exercising.

AI and augmented reality will also power more interactive and immersive news experiences. Readers can explore news stories in a visual and immersive format to experience key events in three dimensions. Imagine an AR news app that plunges users into a 3D simulation of a conflict zone, rolls them through interactive timelines of events, or gives virtual tours of significant locations. These immersions may enhance understanding and induce empathy. But will user preferences ultimately bias these experiences?

## AI News Companions and Their Impact on Society

When everyone can access a powerful AI companion, we expect dramatic transformations in media consumption and equally profound societal consequences. In the worst case, AI companions may amplify today's echo chambers, reinforce existing beliefs, and polarize public opinion further. Social media platforms already use AI to personalize news feeds based on user behavior and increase user engagement. But it also creates filter bubbles, where users get only the news that reinforces their existing beliefs and preferences.

In the best case, AI companions could filter out misinformation and bias and find balanced perspectives when curating news feeds. For example, your AI companion could scan a wide range of sources and highlight different viewpoints so that you can form your own view of complex issues. Could that lead to a more informed public discourse? The potential is there, but how many people will tap it?

Universally available AI companions with advanced data analytics capabilities could also reshape personal privacy, markets, and political dynamics. As a citizen journalist, everyone could scrutinize powerful institutions as thoroughly as the ICIJ did the Panama Papers, though everyone might not have the resources to defend against libel lawsuits. Indeed, with AI companions analyzing petabytes of data, transparency in governance and business could increase significantly. Individuals could use their AI to investigate government actions, corporate practices, and financial transactions in order to hold public and private entities accountable for their actions. We can easily see how these tools could lead to more engaged citizens. Of course, they could also lead to information overload, confusion, conspiracies, and lawsuits.

Universal access to advanced AI companions could democratize economic opportunities by giving individuals tools more powerful than those already available online for financial analysis, market research, and investment strategies. Small businesses and entrepreneurs could use AI to gain insights into market trends, optimize operations, and identify growth opportunities as larger corporations do. Individuals could also better manage their personal finances and make more informed decisions about savings, investments, and expenditures.

Now, consider the impact on trust and reputation. Before you even schedule a business meeting, your AI companion could fact-check statements, verify credentials, and analyze social media content to brief you on the backgrounds, opinions, and behaviors of those you meet with. Such briefings may bias you for or against the people in the meeting. Remember, people do change, for better or worse. Collect your own data and compare notes with other human beings as well as your identic AI.

Then, there is the vital issue of truthfulness, accuracy, and authenticity in the media. We already see how AI can generate realistic but false news stories, images, and videos, so that audiences cannot distinguish between real and fake news. Such content can spread rapidly on social media, causing panic and confusion before someone can debunk it. Indeed, countries have already recognized deepfakes as national security threats.[39] Individuals and organizations could use AI agents to surveil, manipulate, or exploit other individuals or organizations.

In response to such risks, outlets such as the AP have established guidelines for the ethical, transparent, and accountable use of AI in journalism.[40] Under these guidelines, AP alerts readers to its use of AI in generating content and checking the integrity of news.

While universally available AI companions may empower individuals, they also challenge privacy, ethics, and regulation. This new landscape will require a careful balance between leveraging the benefits of advanced AI and safeguarding against misinformation, libel, and other torts. The media industry, including social media, will need codes of conduct. Governments may pass new regulations with stiff penalties for abuse.

Preparing current and future workers for identic AI in the newsroom will also be critical. By automating routine tasks, AI allows journalists to focus on more complex and investigative work to up the overall quality of news. However, as AI takes on tasks like data analysis and news writing, entry-level journalists may have fewer opportunities in traditional organizations. So news organizations must seek a balance between leveraging AI's capabilities and maintaining a human workforce.

By embracing AI's benefits while safeguarding journalism's core values of truth, integrity, and public service, news organizations might remain a trusted and vital resource in our society. The future of journalism is not just about technology; it's about using that technology to inform and serve the public.

# MANAGING THE DARK SIDE
# OF THE AI-ENABLED CREATIVE REVOLUTION

While AI-enabled creativity opens exciting frontiers for self-expression and innovation, it also introduces a complex array of risks and challenges that cut across all creative fields, including art, music, journalism, and film. AI's ability to automate creative tasks threatens employment, traditional business models, fair compensation, and intellectual property rights. The potential overreliance on AI for creative processes may erode human craftsmanship, authenticity, and meaningful storytelling. The increasing use of AI in generating content raises questions about authorship, cultural homogenization, ethnic and gender bias, and the loss of diversity in creative expression.

We have no answers to these profound issues. But we will pose some questions and offer guidelines for shaping a future where AI empowers, rather than undermines, human creativity.

## Building New Business Models for Creative Culture

AI's improving ability to automate creative tasks threatens employment in creative industries. AI systems may replace people in jobs that once relied on human ingenuity, such as graphic design, video editing, and even journalism; AI can perform these tasks faster, cheaper, and more efficiently. While AI may generate new opportunities in fields like AI development or creative direction, the automation of core creative tasks could displace many workers who rely on their artistic skills to earn a living.

The writers and actors who went on strike in 2023 sought to protect their basic rights as creators and prevent studios from using generative AI to reuse and repurpose their completed works.[41] Although the writers reached a tentative deal with the studios, the challenge extends beyond Hollywood. Streaming has drastically reduced earnings for musicians, songwriters, and composers. Meanwhile, the journalism industry has been gutted, with thousands of jobs disappearing. Now, a new wave of AI-powered platforms like Scribe, ChatGPT, and Jasper threatens to disrupt technical writing and corporate content creation, putting even more creative professionals at risk. Even software developers aren't immune as GitHub Copilot becomes more proficient at writing code.

In *Web3: Charting the Internet's Next Economic and Cultural Frontier*, Alex Tapscott argued that culture needs a new business model and that new technologies like AI and blockchain could help.[42] Tapscott explained that we are moving into a world where technology can protect our creativity:

> In this world, technology would make it virtually impossible for anyone to profit from your work without your permission. Instead of making it easier for people and corporations to steal, copy, or monopolize your creative work, technology would empower you to control the destiny of your efforts. A world such as this would recognize and honor the essential value of creativity to human culture. It would protect artists and creators by default instead of putting their livelihoods at risk.[43]

As he points out, Web3 provides the technological framework for this world. In essence, Web3 is the "ownership web," and it remedies many of the internet's shortcomings. The core technologies that will shape the Web3 era include blockchain, artificial intelligence, and extended reality (XR) user experiences. Together, these technologies give creators the tools to create, own, and control their content. For example, with Web3 tools, creators can digitize any asset—a song, photograph, video, voice sample, article, speech, script, joke, their likeness—by placing it in a software container called a *token* protected by cryptography. The asset can be managed and sold without powerful intermediaries like media conglomerates and tech giants.

Arguably, the most significant benefit is that Web3 simplifies tokenization so that creators can more easily monetize and track the usage of their property. If you create art on a Web3 platform, you can secure prompt and fair payment for your work. You can grow and even thrive along with AI, receiving royalties instantly whenever companies use your work to train AI tools and solutions such as LLMs. This has already worked for Hollywood in Web3. Creators made $24 billion selling their work in Web3. By contrast, Spotify paid out $7 billion in royalties to artists last year.[44]

According to Tapscott, we can extend this capability to the metaverse, the web's next cultural and spatial frontier. Facebook's Meta proposes a

Web2 metaverse where a single company owns the world, sets the rules, captures the data, and harvests all the economic benefits. By contrast, open Web3 environments such as The Sandbox are decentralized, with no corporate owner extracting rent and expropriating creators' work. Users can transact freely peer to peer, reaping the reward of their toil and protecting their digital property rights without kicking up fees of 50 percent or more to Big Brother every time value changes hands.[45]

In a Web3 world, all individuals who create value can own and benefit from their contributions. For example, musicians can publish content on platforms like Audius with its 7.5 million users. They not only receive compensation for their creations but also earn an economic stake in the platform's success and a vote in its governance. Similarly, visual artists can bypass galleries and sell directly to fans on platforms like Magic Eden, Rarible, or OpenSea, electing to earn royalties on aftermarket sales. At last count, at least three hundred different creator-led Web3 projects have made at least $1 million in these secondary royalties.[46]

This news is good for creatives on the picket line. Technology can support a business model that benefits all creators by removing gatekeepers, intermediaries, and unnecessary friction. It equips artists with reliable tools to connect more deeply with fans and patrons, protect their IP, and earn a fair and livable share from their work.

## Preserving Human Creativity

Securing the economic future of creative industries is vital. But so, too, is preserving the human creativity that inspires innovation, connects cultures, and captures the beauty in the world around us. Without nurturing the originality and emotional depth that only humans can bring, the creative landscape may lose the authenticity and diversity that make art, storytelling, and design so poignant.

As AI systems learn to create content autonomously, the lines between human creativity and machine-generated work may blur. When AI can generate art, music, films, and books with little to no human input, the question arises: What is original? AI systems rely on huge datasets of existing works to produce new works; the AI-generated content often derives from its training inputs. Artists and creators may feel pressure to compete with machine-generated works by using AI in their own work.

AI companions that handle much of the technical and creative workload may erode human skills in their crafts. As people rely more on AI to perform tasks like composing music, writing articles, or designing visuals, they may lose their edge in these skills. In the long run, AI could create a generation of creatives who lack the fundamental skills and knowledge to produce original work independently. This shift could diminish the role of craftsmanship and mastery in creative fields, reducing human involvement to curation and direction rather than creation itself.

Of course, we asked many of the same questions about synthetic music. For old-school lovers of live music, we miss the days when performance music came from live bands. Today, most clubs feature a DJ creating digital music. The new generation loves it. Perhaps people won't care whether a human or an algorithm generated the art. They'll care only about the experience and the creative content.

Do they care about authorship and ownership? If an AI system generates much creative content, then who owns the final product? Is it the human who provided the input, the developer who built the AI system, or the AI itself? In the United States, for example, the US Copyright Office has determined that human beings may not file for IP ownership of AI-generated works; the works go instead into the public domain.[47] Where ownership rights and royalties are central to creators' livelihoods, how do they earn a living in industries like film, music, and art? When AI-generated content becomes ubiquitous, human creators may struggle to receive credit or compensation for their contributions, marginalizing their role further in the creative economy.

AI systems are only as good as their training. Did their trainers use data representing dominant perspectives and marginalizing minority views? If everyone uses the same or similar models, then AI-generated content could homogenize cultural expression.

Finally, as writers and thinkers ourselves, we fear that creators could rely too heavily on identic AI for inspiration, innovation, and decision-making. The danger here is that creativity becomes algorithmically driven, with AI systems determining what is popular, marketable, or valuable. This reliance on machine-led creativity risks stifling human intuition, spontaneity, and emotional depth, qualities at the core of artistic expression.

## The Path Forward for Creative Professionals

Now is an exciting and terrifying time for creative professionals. Identic AI can empower creatives while undermining their livelihoods. The path ahead may seem murky, but we have some guidelines to navigate the way forward.

The first step is to consider carefully how to balance innovation and authenticity with convenience and skill. With agentic AI, more people—the more diverse, the better—can contribute to art, music, film, and media.

We enjoy creative works because they emphasize human intuition, emotion, and the unique experiences of diverse people. While AI companions can assist in technical tasks, the human touch and human experience should remain central to the creative process. Artists and creators can use AI as a tool to augment their vision rather than replace it, just as we harnessed AI to help research this book, and a filmmaker might turn to AI to enhance the special effects in his or her film. In other words, AI companions should enhance the creative process, allowing creators to experiment with ideas, iterate faster, and explore new creative avenues.

Countering the risk of skill erosion poses an interesting dilemma. Writers, filmmakers, journalists, and other creative professionals will operate in creative fields under AI influence. Training programs in the creative arts should integrate AI tools responsibly so that artists and creators continue to develop their craft alongside AI. Schools, institutions, and creative industries can offer dual training in traditional artistic techniques and AI-powered tools and prepare creators to use both effectively. Investing in lifelong learning can help individuals to adapt.

Transparency is another key principle. We think openness about AI-generated content is the best policy. A good starting point is for creatives and industry leaders to set guidelines and ethical standards for AI-generated content. Consumers should know whether AI created or influenced content. Platforms could include AI content labels that clarify the extent to which AI generated content for films, music, or articles in order to differentiate between machine-generated and human-created works.

Consumers can also help manage the risks of AI-enabled creativity. For example, audiences can demand authenticity and human involvement in the content they consume by being more discerning and aware of the sources of their media. Perhaps platforms should offer filters to screen out AI creations.

As for policymakers, they need to develop new rules and guidelines to address ownership and control over creative works, so that creators who use AI can retain their intellectual property rights. Legal structures must also address the responsibility and accountability of AI-generated content, particularly for misinformation or harmful media outputs. To counter the risk of bias in AI-generated content, creative industries should train AI systems on diverse, representative datasets. Working with underrepresented voices and ensuring that AI-generated content reflects a wide range of perspectives, cultures, and experiences would also help prioritize inclusivity in creative works.

Balancing AI's benefits with the need to preserve human creativity requires concerted efforts from all stakeholders, including creators, consumers, industry professionals, and policymakers. The creative world can leverage AI to enhance innovation without sacrificing the emotional and intellectual depth that defines human artistry. This approach will ensure that agentic AI is an enhancer of creativity, not a replacement for the uniquely human touch that art, music, film, and media require.

## Chapter 7

# SHAPING TOMORROW'S CITIES: AI AGENTS IN URBAN LIFE

I n the heart of the city, Maria steps out of her apartment. Her smartphone—a lifeline provided through a city assistance program—beeps with a notification. It's her AI companion, "Lumi," offering directions to a job interview across town. The AI recommends the fastest public transit route, automatically adjusting for service delays, and reminds Maria to stop by the local daycare center to drop off lunch for her daughter on the way.

As Maria walks to the bus stop, drones buzz overhead, delivering packages and monitoring public safety. Autonomous street sweepers scrape the gum and scoop dog waste from the sidewalks. Inside the bus stop, interactive screens display route updates in multiple languages for immigrants and visitors unfamiliar with the city's layout. Fortunately, AI-driven public transport never keeps her waiting more than a few minutes.

At the community center, where Maria often volunteers, AI companions help residents access essential services: navigating work permits, signing up for childcare programs, and tracking public assistance benefits.

The AI also keeps the center's calendar synced with local events, ensuring families can find affordable activities nearby.

Later that day, Maria joins a workshop on financial literacy. AI tools assist attendees with creating bank accounts and managing digital payments, empowering those without traditional financial resources to participate in the city's economy. The presentation is delivered in real time with AI translation services, breaking down language barriers that might otherwise prevent meaningful engagement.

As evening falls, Maria returns to her housing complex. Lumi senses her anxiety and suggests mindfulness exercises tailored to her mood. It also reminds her of tomorrow's health clinic appointment, where an AI nursing attendant will streamline her care. The city may not have solved every challenge, but it's moving toward a future where technology bridges gaps in access, ensuring no one is left behind.

In this city shaped by identic AI, technology is not just a luxury. It's a lifeline. The city has perfected mass customization: harnessing a vast data fabric to anticipate friction points in residents' lives and remove them, giving each citizen a customized experience regardless of their circumstances. The promise is not only the convenience. It's the potential to transform urban spaces into communities that serve everyone. Maria is the new urban dweller, and she has a superpower living in a city where AI anticipates the needs of its most vulnerable residents and works to make everyday life more livable, equitable, and inclusive.

This fictional account of Maria's life portrays a city where traffic jams are a relic of the past, intelligent systems seamlessly orchestrate your daily commute, and public services anticipate your needs before you even know you have them. It may seem like a fantasy for many city dwellers today. But in the world's most cutting-edge urban environments, this vision is fast becoming a reality in what we call the *cognitive city*.

A cognitive city represents a paradigm shift where the mantra is "code over cranes." While physical infrastructure such as buildings, roads, and utilities remains the foundation of any city, the true leap in urban experiences will not come from constructing taller skyscrapers. Instead, it will emerge from the intelligence derived from how we interact with these existing assets, and each other.

Think of it this way: Traditional cities grow outward and upward with cranes, pouring concrete to build the tangible. Cognitive cities, by contrast, grow inward and smarter with code, embedding intelligence into every interaction. It's not the structure of a hospital that will redefine healthcare but the seamless integration of data insights that predict patient needs, optimize care, and enhance outcomes. It's not the design of a park that will elevate urban living but AI companions that personalize recreational spaces.

The greatest opportunities for urban renewal lie not in creating more physical spaces but in unlocking the untapped potential of those we already have. Data becomes the connective tissue that transforms infrastructure from static to dynamic, from reactive to predictive. In a cognitive city, every building, every streetlight, and every interaction is a data point that contributes to a living, learning ecosystem where experiences are continuously optimized, not through steel and concrete but through algorithms and insights.

In essence, cognitive cities prioritize smart utilization over mere expansion. Identic AI will be at the heart of the cognitive city movement, helping urban dwellers, business owners, and city officials make our urban environments more efficient, responsive, and livable.

## COGNITIVE CITIES AND THE NEW DIGITAL INFRASTRUCTURE

If today's cities invest in networks, data, and revamping antiquated processes, they have a shot at building an urban infrastructure that will eventually support everything from intelligent energy grids and real-time traffic management to public health and housing and personalized public services. Unfortunately, most of today's cities lack the technological foundation for transformations this chapter envisions.

Throughout history, human beings have founded stable, prosperous, and inventive societies on public infrastructure. Says Douglas Heintzman of the Blockchain Research Institute, "The basis of infrastructure is technology. From the roads and water aqueducts of ancient Greece and Rome to the systems for trade, commerce, sanitation, and gas distribution of the agrarian feudal city-states and to the electrical power grid, highways, bridges, and telecommunications networks of the Industrial Revolution, infrastructure has enabled our world."[1] However, to meet the requirements of the digital age, our cities need digital infrastructure. The technologies supporting most of our emerging digital economy—data networks, call centers, cloud computing, traffic systems, and financial management—are primitive and limited in what they can support. What's more, private entities built and own them. As a result, the infrastructure of cities at the forefront of the Industrial Revolution no longer meets the rapidly changing economic and social needs of modern urban dwellers. Meanwhile, their administrators struggle to fund or raise funding for big development projects.

What infrastructure do the digital-age cities need to support all the technologies behind identic AI as well as Web3, the Internet of Things, and extended reality? For starters, digital-age infrastructure needs to evolve far beyond roads and bridges. We're talking next-gen connectivity: ultra-fast 5G and fiber networks, edge computing hubs to process data locally in real time, and secure blockchain systems to support decentralized services. Smart grids and IoT sensors must cover everything from traffic lights to trash bins, while extended reality platforms require powerful GPUs and cloud infrastructure to deliver immersive experiences. It's a tech ecosystem that's not just built for efficiency but for empowering people to live, work, and play in ways we're only beginning to imagine.

With such infrastructure in place, residents, businesses, and government agencies could tokenize all their assets, from money and securities to intellectual property, art, votes, and even physical assets like construction materials in a new building. All those groups could trade things of all kinds in a city: not just units of energy but carbon credits. It would facilitate smart contracts so that people and organizations could rapidly make self-executing agreements.

This infrastructure would support not only trillions of smart communicating devices in managing the public and private assets but also millions of citizens communicating, organizing, and voting in democratic elections. As a municipal resource, it would help manage and monitor air, water, renewable energy, sanitation and recycling, law enforcement, public finances, and community safety. Perhaps most important, it would be a cognitive city that learns with its inhabitants in real time and uses its resources responsibly, sustainably, and cost effectively.

Cities like Singapore and Helsinki already leverage AI to manage their water resources more efficiently. AI distributes it, minimizes waste, and uses predictive analytics to maintain pipelines.[2] These intelligent systems will increase the security and reliability of critical infrastructure and the livability of urban environments.

Smart mobility is another cornerstone of the digital infrastructure for cognitive cities, offering innovative alternatives to traditional systems like subways and streetcars. Instead of maintaining costly, fixed networks built in the 1900s, cities could use identic AI to configure ad hoc public transit systems. These systems might include smart vehicles that could reroute to avoid congestion or autonomous energy-efficient shuttles that transport people on demand. Identic AI could tailor mobility services to the needs of individuals and communities. Picture AI agents coordinating grocery pickups and dry-cleaning drop-offs with autonomous shuttles. In Helsinki and San Francisco, pilot projects have demonstrated the potential of autonomous shuttles as reliable, scalable public transportation options.[3]

In a cognitive city, AI agents could streamline bureaucratic processes in city hall so that residents could more easily access municipal services and complete paperwork. AI companions like those in Barcelona help citizens navigate such services, schedule appointments, and report issues.[4] Chatbots will handle routine inquiries and free human public servants to focus on complex tasks. These innovations will improve not only efficiency but the quality of service delivery.

Identic AI could also reinvigorate civic engagement. An AI platform could aggregate input from personal agents to give city officials real-time insights into residents' concerns and priorities. It could also track the commitments and decisions those officials make, the rates of delivering on those commitments, and the outcomes of their decisions. Transparency is essential. Early glimpses include Boston's CityScore dashboard, which uses data analytics to gauge the effectiveness of city services.[5] On AI-driven participatory platforms, residents could contribute ideas and feedback on urban projects. These tools will empower citizens to participate directly in city planning and investment.

Cognitive cities are built on the power of platform economics, where value grows through participation. The true secret of a cognitive city lies in its ability to foster unprecedented levels of engagement among its residents, businesses, and systems. The more people participate, sharing data, insights, and interactions, the greater the collective intelligence of the city becomes. This creates a virtuous cycle: Increased participation drives greater intelligence, and greater intelligence generates more personalized and valuable experiences for every individual. In a cognitive city, the whole truly becomes greater than the sum of its parts, unlocking a new era of shared prosperity and innovation.

While controversial, cognitive cities will also revolutionize public safety through advanced surveillance and emergency response systems. Our AI-powered personal agents will interact with a ubiquitous network of cameras and sensors to monitor public spaces for unusual activities, quickly identify potential threats to our safety, and alert authorities if needed to avert harm. Autonomous AI agents could also assist in disaster management by predicting catastrophes, coordinating rescue operations, and delivering real-time information to emergency responders. Citizens of every jurisdiction could vote on the level of surveillance they want their communities to maintain and could instruct their AI companions to divulge different private information to authorities under different crisis scenarios.

To be sure, integrating identic AI into urban life will come with challenges. City planners and technology providers must address privacy, data security, and equity issues so that citizens share AI's benefits broadly and fairly. As New York City Mayor Eric Adams put it while launching the city's AI Action Plan in 2023, "While artificial intelligence presents a once-in-a-generation opportunity to more effectively deliver for New Yorkers, we must be clear-eyed about the potential pitfalls and associated risks these technologies present."[6] Like many mayors worldwide, Adams is keen to embrace AI in transforming urban life but recognizes the importance of governance structures to safeguard the public interest.

Sundar Pichai, CEO of Alphabet, recently called AI "the most important thing humanity has ever worked on—more profound than fire or electricity or anything that we've done in the past."[7] This will be the case for today's cities. By optimizing services, infrastructure, mobility, and civic engagement, identic AI will improve the quality of life for city residents as long as cities harness its potential in ways that include everybody, enhance everybody's well-being, and promote sustainable urban development. Cognitive cities are not just about chatbots, ubiquitous sensors, and augmented reality but also about creating environments where all people can thrive.

# URBANIZATION AND THE SMART CITY IMPERATIVE

You may be wondering why this shift to cognitive cities is so important. The answer boils down to DEE: demographics, economics, and the environment.

Let's start with demographics. Over the past century, people have sought better opportunities, infrastructure, healthcare, and education in cities. By 2050, nearly 70 percent of the world's population will live in urban areas. Under continued urbanization, new megacities will emerge, and existing ones will expand.[8]

As cities grow, the challenges of managing their infrastructure, delivering services, reducing carbon emissions, and developing their economies also grow. For example, with urban expansion, the demands on public transportation, water supply, energy grids, and waste management intensify, potentially leading to congestion, overloads, longer lags and downtimes, and increased maintenance costs.[9] Cities must continuously invest in upgrading and expanding their infrastructure with sustainable technologies to accommodate growth.

Greater energy consumption and industrial activity increase carbon emissions and other hazards like smog. Cities must transition to renewable energy sources, promote energy-efficient buildings, expand public transit options, and encourage sustainable urban planning practices.[10] Without proactive measures, the environmental impact of urban growth can undermine global efforts to combat climate change. Cities that successfully implement carbon-reduction strategies will contribute to global sustainability and position themselves as leaders in green innovation, attracting environmentally conscious businesses and residents.[11]

Urban population growth also places new burdens on essential services such as healthcare, education, public safety, and social services. Larger, denser populations require more resources and better coordination to deliver these services effectively. Growing cities may face shortages of healthcare facilities, schools, and emergency response units, leading to longer wait times, reduced access, and potential declines in service quality.[12] The increased diversity within expanding cities also demands more inclusive and culturally competent services. Cities must innovate and adapt their service delivery models to meet the evolving needs of their residents while maintaining high standards of quality and accessibility.

Finally, cities have traditionally anchored economic growth to physical infrastructure, expanding their tax base through land acquisition, housing development, and commercial spaces. Cognitive cities, however, redefine expansion by focusing on data and its transformative power to enhance real-life experiences. In a cognitive city, citizenship is no longer tied to residency but to contribution. These cognitive citizens interact with the city's advanced data platforms, leveraging insights to create innovative solutions that improve functionality, sustainability, and quality of life.

By focusing on data-driven innovation rather than physical expansion, cognitive cities can harness the creativity and expertise of their citizens to develop applications and services that make existing assets like buildings, utilities, and public transportation systems smarter and more efficient. These innovations, available through the city's app store, establish a collaborative revenue model: The city benefits financially while cognitive citizens earn income from their contributions. This shift not only drives economic growth but also aligns with global sustainability goals, demonstrating that data can be a powerful tool to address environmental challenges.

Cities that lead in AI also position themselves to attract businesses and top talent. A 2023 report by McKinsey Global Institute projected that AI could contribute up to $2.9 trillion to cities' economies by 2030, underscoring the immense opportunities for the early movers in AI adoption.[13]

AI-enabled urban infrastructure and services will also bring considerable benefits, including enhanced efficiency and reduced operational costs from integrating AI into traffic management to waste collection. A recent Deloitte study highlighted that smart city technologies, including AI, could reduce commuting times by 15 to 20 percent, translating into billions of dollars in increased productivity annually.[14]

As citizens benefit from intelligent services in progressive cities, their expectations of AI-driven conveniences continue to rise. Cities that meet these expectations can enhance the quality of life for current residents and gain an advantage in attracting new ones. A 2019 study by Capgemini revealed that 58 percent of citizens expected AI to improve public services and elevate their overall living experience, positioning forward-thinking cities as desirable places to live.[15]

# HOW AI COMPANIONS WILL RESHAPE CITY SERVICES AND OVERSIGHT

Imagine strolling through your city with your pocket AI companion that anticipates your needs, streamlines your interactions with municipal services, and enhances your recreation. This is the future we are hurtling toward, where autonomous AI companions transform how we engage with local services and our urban environments. By managing taxes, permits, public health information, and other municipal services, AI companions could revolutionize city living.

## Streamlining Municipal Services

Over the last hundred years, municipal bureaucracies have served up lengthy forms, long queues, and labyrinthine procedures for the simplest tasks. AI companions may significantly reduce this complexity. An AI companion like Siri or Google Assistant trained to navigate city services could challenge your recent municipal property tax assessment. Your AI agent can complete the process step-by-step, from filling out and filing forms to following up with the appropriate office.

Of course, most cities are just beginning to harness identic AI technology. For example, under Mayor Adams, New York City is experimenting with AI to improve local services. The city's AI chatbot helps residents find information on a wide range of municipal services, from trash collection to noise complaints. This chatbot responds immediately, reducing the wait times and workloads of human operators. But the service is immature. In April 2024, a nonprofit newsroom tested the chatbot, which told employers that they could take a cut of their workers' tips and landlords that they could discriminate by sources of income.[16]

Mayor Adams insisted the city would "iron out the kinks," and NYC's missteps have not deterred other cities from following suit.[17] Singapore, for example, has a chatbot named Jamie that helps answer resident queries.[18]

Jamie can handle complex queries by asking follow-up questions to refine its responses.

Both NYC and Singapore give us a glimpse of identic AI cutting through municipal red tape. Soon, a small business owner looking to open a new café or a resident wanting to renovate their home will ask their AI agents to gather required documents, fill out the forms, keep track of the application status, and notify them of any additional requirements.

Streamlining municipal services may be step one. Step two may be reinventing the modus operandi of government. When Joseph was an executive on an Internet of Everything project, his team focused on connecting retail environments to improve customer experience. One of the project's key goals was to reduce customer wait times by tracking the items in shopping carts so that the retailer could staff checkouts more efficiently. During trials, the retailer reduced wait times by about 30 percent, a significant achievement.

Joseph shared this achievement with his daughter, around eleven years old at the time. She said, "That's cool, Dad. I would have done something a little different." "Like what?" Joseph asked. She replied, "Instead of reducing the lines, why not get rid of them?"

Her suggestion prompted Joseph to rethink the entire approach. The project team began exploring how AI could eliminate lines altogether. This shift in perspective inspired the team's first mobile checkout prototype, where customers could check out without waiting.

In the age of AI, we must mind not only how we're training AI—the data and the model we're using—but also what we're asking of AI so that we unlock the full potential of the technology.

## Transforming Urban Recreation and Public Health

Beyond streamlining services, AI companions can enhance tourism in urban settings. Picture a day at the park with an AI companion that suggests the best walking routes based on real-time weather and crowd data and shares insights into the landmarks you pass. For a weekend outing, your AI companion recommends events, books tickets, and personalizes your itinerary. This level of curation makes exploring every city more engaging.

The COVID-19 pandemic underscored the importance of timely, accurate, and comprehensible public health data. AI companions can help visualize, disseminate, and map information on vaccination sites, health advisories, and local infection rates. This proactive approach could help minimize spreading and save lives, especially in dense neighborhoods and medical deserts.

## Enhancing Civic Participation

In *Macrowikinomics*, Don Tapscott and Anthony Williams argued that the proliferation of digital technologies and collaborative platforms marked the second era of democracy, characterized by active and meaningful citizen participation in governance.[19] Don and Anthony envisioned a transition from hierarchical, top-down systems of governance to more inclusive, participatory models. In this second era of democracy, they saw citizens as active contributors to public policies, shaping decisions through open data, crowdsourcing, and digital platforms for direct dialogue between public servants and their constituents.

The second era of democracy did not fully materialize. Perhaps with identic AI infrastructure, citizens will take active roles in government. AI companions can transform civic engagement by personalizing government. In the traditional model, citizens must invest time and effort to inform themselves and engage in local governance. But with AI companions, citizens could receive real-time and relevant updates. Some cities already use platforms like IBM's Watson to analyze large datasets and generate insights to guide public policy. Ordinary citizens could have a personal AI analyst who breaks down the complexities of urban planning, environmental policies, or educational reforms so that they can advocate and debate effectively.

Integrated with local government databases, Apple's Siri or Amazon's Alexa could notify us of the agenda of each upcoming town hall meeting, public hearing, and community event. It could remind us to vote, pro- vide information on candidates and issues, and facilitate discussions with neighbors.

## Shaping Public Policy and Increasing Accountability

The potential of AI companions extends to shaping public policy. Through continuous interaction with citizens, AI could aggregate public opinion for policymakers. An AI companion could collect and analyze user senti- ments on various issues, identifying common concerns and suggestions. Citizens could share their aggregated data with their representatives, help- ing them to understand the needs and priorities of their constituents more accurately. Cities like Boston and Barcelona are already experimenting with AI-driven platforms to gauge public sentiment and improve decision- making. By acting as intermediaries, AI companions could amplify citi- zens' voices in policy-making.

Holding elected officials accountable is a cornerstone of democracy. AI companions could track the campaign promises and in-office actions of officials and generate a live dashboard of their performance on their commitments.[20] Furthermore, AI agent-driven transparency tools could highlight instances of corruption or inefficiency, giving citizens evidence for action.

By making information more accessible and actionable, AI compan- ions can strengthen democratic oversight and governance. But only if cit- ies and technology companies help bridge the digital divide and quickly. Otherwise, AI may widen it, perhaps polarizing citizens even further.

Weaving autonomous AI agents into the fabric of local democracy carries other significant risks, such as corporate or foreign actors exploit- ing AI agents to reengineer public sentiment. For instance, AI-driven platforms could selectively amplify consenting voices while marginalizing dissent, steering public discourse in favor of outcomes that benefit those in power.

Then there are the bots masquerading as humans in online forums, social media, and public comment platforms. Increasingly sophisticated agents could participate in debates, share opinions, and even appear emotionally engaged so users cannot distinguish between real humans and AI entities. This so-called *astroturfing*—the practice of creating fake grassroots movements—could make it appear as though a large segment of the population supports an issue when, in reality, it's automated agents.

In short, the future of civic engagement requires more than passive optimism. It demands vigilance, transparency, and accountability in how we build and govern identic systems. The true promise of cognitive cities isn't just in making urban life more efficient; it's in rebuilding trust, strengthening civic bonds, and ensuring technology serves everyone. If we actively manage the dark side, then integrating autonomous AI agents into local services and democracy could positively transform urban living. The cognitive city experience will go beyond convenience to genuine community engagement.

# THE FUTURE OF TRANSPORTATION: HOW AUTONOMOUS AGENTS WILL TRANSFORM MOBILITY

Smart urban mobility may be one of the most transformative aspects of urban infrastructure in the age of autonomous AI agents. While AI could enhance city management, such as optimizing water flows, energy grids, and waste management, the revolution in urban mobility stands out for its profound and immediate impact. Through self-driving vehicles, intelligent traffic management systems, and dynamic public transportation networks, autonomous AI agents could redefine how people and goods move within cities, reducing traffic congestion, commute times, and carbon emissions while improving safety, efficiency, and resilience in transportation ecosystems.

Self-driving cars and public transport vehicles are good examples of autonomous AI agents in action. In essence, we are talking about intelligent mobility services designed to move city residents from point A to point B in the most efficient, timely, and sustainable way possible. By integrating real-time data and predictive analytics, these AI-driven mobility solutions could respond dynamically to changing conditions, offering tailored routes and services that align with the needs of urban dwellers. Let's trace the remarkable evolution of self-driving cars and some challenges in bringing this technology to fruition.

## The Rise of Autonomous Vehicles

The dream of fully autonomous and intelligent transportation has long captured the imagination of technologists and science-fiction enthusiasts. In a perfect world, we could turn our garage into a spare bedroom while our self-driving car is out making money as an Uber. It could increase the mobility and autonomy of the blind and protect sober drivers from inebriated ones.

We can trace the early stages of autonomous transportation back to the 1920s when innovators experimented with radio-controlled vehicles. In the 1980s, research institutions and automotive companies began collaborating more seriously on concepts of autonomous vehicles.[21] They focused on developing advanced sensors, computer vision systems, and control algorithms through which vehicles would perceive and move safely within their surroundings. AVs debuted on closed tracks and controlled environments so that technologists could experiment with their capabilities.

The 2000s marked a turning point as advancements in computing power, sensor technology, and machine learning algorithms propelled autonomous transportation to new heights. The Defense Advanced Research Projects Agency (DARPA) organized the DARPA Grand Challenges, which spurred competition among researchers to develop fully autonomous vehicles that could complete challenging off-road courses.[22]

Major automotive companies also entered the race, investing heavily in AV R&D. They focused on refining sensor technologies such as lidar (light detection and ranging), radar, and cameras, as well as developing sophisticated AI algorithms to process and interpret sensor data in real time.

By the 2010s, AVs started hitting public roads, albeit in limited testing environments and under strict supervision. Companies like Google's Waymo, Uber, and Tesla gained significant attention for their efforts in advancing autonomous driving technology. The industry rapidly expanded testing programs in various cities worldwide, accumulating valuable data and insights.

Waymo, a subsidiary of Alphabet, has been at the forefront of AV technology. With millions of miles driven on public roads and in simulations, Waymo's self-driving cars have demonstrated impressive safety records.[23] By analyzing vast amounts of data in real time, these vehicles can predict and avoid potential hazards, making roads safer for everyone. Indeed, one of the most significant benefits of AVs is higher street safety. Without fatigue, intoxicants, or distractions, AI-driven systems can react faster than human drivers.

A full-scale rollout of AVs could significantly reduce accidents and fatalities. According to the National Highway Traffic Safety Administration, human errors cause 94 percent of serious crashes.[24] Waymo's report on its safety performance indicates that its AVs reduced potential crash scenarios by over 90 percent compared to human-driven cars in the same conditions.[25] Research by the RAND Corporation predicts that widespread adoption of AVs could prevent tens of thousands of deaths annually in America alone.[26] Meanwhile, McKinsey & Company has projected that a significant reduction in accidents would save $190 billion annually in the United States by reducing accident-related costs, including healthcare, vehicle repairs, and lost productivity.[27]

Autonomous vehicles could also improve accessibility and autonomy for older adults and individuals with disabilities. As safe and reliable transportation options, AVs can enhance mobility and independence for those with difficulty driving, especially at night or in poor weather. Companies like Ford and Toyota are exploring inclusive design features for their AVs, such as wheelchair accessibility and voice-activated controls so that their AVs cater to diverse needs in transportation.

Finally, AVs have the potential to reduce the environmental impact of personal transportation. AI can optimize driving patterns for fuel efficiency, and the widespread adoption of electric AVs can decrease greenhouse gas emissions. Rivian, an electric vehicle manufacturer, is designing a fleet of electric AVs to minimize environmental impact through zero-emission technology and AI-driven efficiencies. As more companies adopt similar approaches, we can expect greener transportation.

## Navigating the Challenges

Despite the potential, the automotive industry has delayed fully implementing AVs. For starters, autonomous systems can operate safely in structured environments but less so in unpredictable scenarios and edge cases, as when a human driver knocked a pedestrian into the path of a Cruise AV, which then dragged the person a few feet, seriously injuring her.[28] Training AI to anticipate rapidly changing road conditions, fickle human behavior, and extreme weather events remains a complex task. So far, Google and others have spent over $100 billion on it, yet they are years off.[29]

Other challenges are navigating through dense urban environments with intricate traffic patterns, pedestrians, cyclists, and various road users, and maintaining detailed maps of roads with lane markings, traffic signs, and temporary construction zones. Simple traffic cones can confuse AVs.[30] If a prankster drew whitewash lines veering off the road into a tree, a human driver would likely recognize it as a prank and stay on the road, but an AV might not.

Above all, AVs pose ethical dilemmas. For example, autonomous cars may confront traffic situations where they must choose one of several options, all of which could harm humans.[31] How should AVs prioritize human life? Without resolving such a fundamental concern, industry leaders could fail to earn the public's trust that such innovations need for mainstream adoption in society.

Given these limitations, the focus in these early days of developing autonomous technology has shifted toward specific use cases, such as ride-sharing services, delivery vehicles, and public transportation. Companies like Waymo and the French firm Navya have launched pilot programs and commercial services, bringing autonomous mass transportation closer to everyday life. Navya's autonomous shuttle operates in comparatively controlled urban environments worldwide, delivering efficient, reliable, and cost-effective mass transit services with more flexible routes and schedules.[32]

By specializing in delivery vehicles and public transportation, developers can work within controlled environments and scenarios in well-mapped urban areas or during specific operating hours. These specific use cases help the industry to refine the technology, cultivate public and policymaker trust, and build out the regulatory and economic infrastructure for mass adoption.

## Mobility as an Experience

Let's assume that rapid advances in autonomous driving technology will help vehicle manufacturers, regulators, and other stakeholders over the speed bumps. The next step will be harnessing autonomous vehicles to transform mobility into a experience. After all, the true potential of autonomous vehicles goes far beyond just transportation. Just as cell phones evolved from being devices for making calls to becoming essential tools for texting, photography, gaming, and much more, autonomous vehicles are poised to redefine what a "car" can be in our daily lives.

In the future, cars will become multifunctional spaces, tailored to the needs of their users. Imagine autonomous vehicles doubling as mobile offices, where professionals can hold virtual meetings, collaborate in real time, or even access on-the-go coworking pods. Families might transform their vehicles into entertainment hubs with immersive VR experiences, offering games, movies, or virtual museum tours while on the road. For those seeking relaxation, AVs could provide spa-like environments with massage chairs, ambient lighting, and calming music. These vehicles will no longer simply move us from point A to B; they will actively enrich the journey itself.

Cars will no longer just be vehicles but will become destinations in their own right. Urban commuters might book shared AV pods equipped with fitness machines for a morning workout on the way to work. Shoppers could step into retail-branded AVs, offering curated product showcases or augmented reality fitting rooms. Dining experiences could evolve, with food trucks replaced by gourmet-restaurant AVs that bring a meal and a chef directly to your location. For long-distance travelers, luxury AVs could function as rolling hotel suites, complete with beds, minibars, and panoramic windows for scenic routes.

The daily grind of commuting will never be the same. When combined with AI-enabled traffic routing, your AV will provide users with real-time, tailored navigation recommendations based on their unique preferences, schedules, and current city conditions. AI companions will also pave the way for a new transportation economy where commuters can seamlessly purchase goods and services as part of their daily journeys. As commuters move through the city, their AI companions can anticipate their needs and automatically coordinate these tasks within their transportation plans. For instance, if a commuter typically buys a coffee during their morning commute, the AI companion could preorder the coffee from a café along the route, making the payment automatically, and schedule a brief stop. Similarly, if a commuter needs to pick up groceries, the AI companion could arrange for delivery to a convenient location on the commuter's route.

This integration would create a more fluid and efficient transportation economy where residents use their commuting time productively. As AI companions learn from the commuter's behavior, they can increasingly personalize services, blurring the boundaries between commuting and consuming. In this on-the-go economy, transportation is not just about getting from point A to point B but also about enhancing the quality of daily life.

## THE AI GUARDIANS: REVOLUTIONIZING PUBLIC SAFETY IN URBAN LANDSCAPES

The practice of surveillance has always been controversial, a tool for securing public safety at the expense of individual privacy. Autonomous AI agents may flip this script, helping individuals to monitor and analyze their urban environments. Picture a network of AI-powered cameras and sensors strategically placed throughout a city, capable of identifying suspicious activities, detecting criminal behavior, and predicting potential threats. These systems can analyze vast amounts of data in real time, recognizing patterns that human eyes might miss. Now patch your personal safety agent into this smart surveillance network to create what we call secure serenity: a world where AI systems create an invisible shield around every individual, dynamically adapting to their surroundings and anticipating threats before they materialize.

This isn't just about cameras or sensors; it's about an integrated intelligence that ensures safety as an ambient, ever-present reality. From autonomous drones patrolling public spaces to AI companions that discreetly guide individuals away from potential danger, secure serenity allows life to be lived without the mental bandwidth consumed by personal safety concerns. In such a city, parents never have to worry about their children's walks to school, travelers can explore with unbridled freedom, and neighborhoods thrive as trust replaces fear.

Let's say a tourist is returning to their hotel late at night. The tourist's personal AI-powered urban safety agent uses real-time data from the city's network of AI-driven cameras, sensors, and other connected devices, combined with GPS data, to monitor the safety of each block. The AI agent analyzes local crime patterns, real-time security data (such as crowd movement and traffic flow), and potential threats based on prior incidents and steers the tourist clear of risky blocks. If the AI agent detects a group of individuals behaving suspiciously in a nearby alley, it can alert the tourist and suggest an alternative, safer route.

The agent could monitor environmental changes, such as rapidly changing weather, dense crowds, or ground tremors, and notify the user of potential hazards. For example, it could provide real-time warnings in emergencies, such as a fire or a public disturbance in progress. In immediate danger, the AI agent could initiate emergency procedures, such as calling local law enforcement, notifying the nearest surveillance cameras, or alerting nearby security personnel. The agent could also trigger distress signals and share real-time location data with authorities or trusted contacts to get a rapid response.

With an AI companion constantly monitoring their environment, people would feel more secure, especially in high-risk areas or during vulnerable times like late-night commutes. The AI agent's ability to preemptively detect and avoid threats could lead to fewer incidents of crime or injury. This AI-enabled urban safety model could significantly alter how cities protect residents and visitors, making safety more personalized, adaptive, and efficient. That said, it also requires careful consideration of ethical, legal, and security frameworks.

Critics may argue that secure serenity infringes on individual privacy, but this perspective overlooks a critical truth: Privacy is inherently contextual, and its value is directly proportional to the benefits one receives in exchange. In the cognitive city, where personal safety increases significantly, the value proposition is clear. The assurance of life lived free from fear, where safety is ambient and invisible, reframes the debate about privacy.

Of course, constantly monitoring individuals' locations and behaviors will still raise concerns. So, to protect residents' privacy, cities would need to establish strict regulations around data collection, usage, and sharing. AI systems that rely on predictive policing data could also inadvertently reinforce biases present in data, leading to unequal protection for residents based on factors such as race or socioeconomic status. Then, hackers could harness AI agent technology to perpetrate more sophisticated crimes. Indeed, as cities and citizens integrate AI into their systems and personal devices, they open digital pathways that criminals might exploit.

## AI-Powered Urban Criminality

Criminals are typically among the first to embrace new technologies, from the automobile and the handgun to smartphones and cryptocurrencies. Criminals may already be using AI agents to manipulate or evade surveillance systems. They might deploy AI to create deepfakes that confuse or mislead security systems or to jam or manipulate signals from AI-based surveillance cameras, sensors, or drones. AI agents could also be used to identify blind spots in these systems, helping criminals navigate cities undetected.

Criminals could use AI agents to automate logistics for illegal operations, such as coordinating drop points for drugs, stolen goods, or weapons using AI-enhanced navigation systems and self-driving vehicles. These systems could optimize routes to avoid detection or police patrols, making criminal networks more efficient and difficult to disrupt. Furthermore, they could program autonomous drones to deliver contraband or conduct reconnaissance for illicit activities while avoiding surveillance.

By analyzing police patrol schedules, public safety responses, and other data, an AI crime agent could help criminals choose ideal times to strike or find new methods of attack that exploit weaknesses in urban security networks.

Social engineering scams are also a concern. AI agents might generate highly convincing messages and impersonate voices or texts, allowing criminals to conduct more sophisticated phishing or fraud schemes. AI-driven agents could mimic trusted sources, simulate the behavior of real people, or generate disinformation. In short, while AI agents offer significant opportunities for improving urban safety, criminals could leverage the same technology to carry out more sophisticated crimes.

## Law Enforcement in the Age of AI Agents

Of course, if law-abiding residents and criminals can have AI agents, then so, too, must the police. Indeed, the potential criminal misuse of AI underscores the importance of ethical AI development and robust security measures. Cities will need to invest in counter-AI technologies and strengthen cybersecurity defenses to anticipate and mitigate these threats. Policymakers will also need to establish regulations governing AI use so that its deployment for safety doesn't generate vulnerabilities.

Consider the City of London, one of the most surveilled areas in the world.[33] With an extensive network of CCTV cameras—estimated to be over 600,000 cameras or nearly 70 cameras per 1,000 residents—the city has long used technology for public safety.[34] The city has placed these cameras strategically in commercial districts around Oxford Street and Piccadilly Circus, as well as residential neighborhoods and transport hubs.

Combining this comprehensive surveillance system with the power of AI agents could provide a model for other urban centers aiming to enhance security through technology.

Today, law enforcement officials monitor the footage. Integrating AI agents into this extensive CCTV network could revolutionize how cities and police forces manage public safety.

The City of London already uses AI-powered facial recognition technologies to enable faster identification of suspects and tracking of suspicious individuals.[35] Now, imagine an AI agent with the power to not only cross-reference faces with criminal databases but also crawl the web to pull in data from other sources, including social media posts. For example, law enforcement can instruct their AI agents to analyze social media activity for indicators of gang violence, drug trafficking, or other illicit activities. By recognizing these patterns early, law enforcement can intervene before crimes escalate.

Integrating AI agents with London's CCTV system could also enhance operational efficiency. By automating routine monitoring tasks, AI agents could free up human resources to focus on more complex and strategic aspects of public safety. Redeploying personnel could lead to faster resolution of incidents and a more efficient allocation of police, fire, and sanitation resources. Say a large protest erupts in a public square or a fire in a public building; AI agents could predict crowd movements and identify potential risks, thus helping the police and fire personnel to manage the crowds and protect public safety. In airports and public venues, AI agents could help detect weapons, explosives, or other contraband through enhanced scanning and data analysis.

When prosecuting crimes, investigators could deploy AI agents in forensic investigations to analyze evidence faster and more accurately. Picture an episode of *Criminal Minds*, where the "lab technicians" are advanced AI agents capable of processing terabytes of data in seconds, reconstructing crime scenes in virtual reality, and identifying patterns or connections that human eyes might miss. An AI agent might assist in DNA analysis, helping to match samples more quickly and detect relationships between different cases. In cybercrime investigations, AI agents can comb through massive amounts of digital data, detecting anomalies that may indicate hacking, identity theft, or fraud.

## Urban Paradise or Orwellian Panopticon

While the potential benefits of AI in public safety are significant, we cannot ignore the ethical implications. On one hand, these cities could become utopias, enhancing quality of life, fostering inclusivity, and empowering citizens through innovation. On the other hand, they could mirror the dystopia depicted in George Orwell's *1984*, where pervasive surveillance, data misuse, and authoritarian control erode privacy and freedom. The outcome hinges on how these technologies are implemented and governed, and whether they prioritize humanity and well-being over control and power.

The scale and reach of London's surveillance system, for example, have made it an effective tool in combating crime, but it also raises privacy concerns. Civil liberties groups have voiced concerns about the potential for misuse, particularly with facial recognition, which can result in the surveillance of innocent individuals and the amplification of biases in policing. In highly monitored areas, individuals could be under constant scrutiny without consent, even in situations where there is no legal justification for it. This overreach could lead to a society that compromises personal freedom due to the pervasive presence of government watchfulness.

Governments could misuse agents for mass surveillance purposes, monitoring citizens on a large scale without transparency or oversight. It is easy to see how authoritarian regimes would use such systems to suppress dissent, monitor political opponents, or target activists. For instance, human rights advocates have criticized China for using AI and facial recognition to monitor ethnic minorities and maintain social control, particularly with its surveillance of Uyghur Muslims in Xinjiang.[36] However, can we trust democratic governments to refrain from deploying similar practices?

AI agents used in surveillance systems can inherit biases from training data, leading to discriminatory profiling. If developers train the system on biased data, it may unfairly target specific racial or ethnic groups, resulting in higher rates of false positives. Studies have shown that facial recognition systems are more prone to errors when identifying people of color and women, raising concerns about the fairness and equity of these systems in public safety.[37]

The use of AI-enabled surveillance could also discourage citizens from exercising their rights to free speech and public assembly. Knowing that officials are constantly monitoring their activities may deter people from participating in protests, demonstrations, or other forms of public discourse that are essential for a functioning democracy. This chilling effect on free speech is a significant concern in highly surveilled environments.

There is also abundant potential for misuse and exploitation. AI-powered surveillance systems collect vast amounts of data, making them prime targets for cyberattacks. If hackers gain access to these systems, they could steal sensitive personal information or manipulate surveillance data for malicious purposes. Corporations or rogue bureaucrats could misuse surveillance data for purposes unrelated to public safety, such as monitoring consumer behavior or targeting political opponents.

With detailed surveillance data, individuals could become targets for harassment, blackmail, or other forms of exploitation. If an AI agent can track personal habits, movements, and interactions, it could easily profile individuals for malicious actors to exploit for personal gain. The potential for misuse—ranging from privacy invasions and biased policing to government overreach—necessitates the implementation of stringent legal frameworks, transparency, and ethical oversight to mitigate these risks.

Ultimately, the balance between security and freedom is delicate and often contentious. We can use identic AI to enhance public safety without infringing on individual rights. Cities must establish policies and regulations to ensure that relevant agencies are ethical and responsible in their use of surveillance and emergency response technologies. Policies should include clear guidelines on data retention, access controls, and accountability mechanisms to address any misuse of power.

The integration of identic AI will shape the future of public safety in large cities. Autonomous agents will likely work silently and efficiently to protect residents as part of urban infrastructure. As these technologies evolve, cities must continuously assess and refine them to address emerging challenges and ethical considerations. The goal is to create a safe, secure, and equitable urban environment where we harness the benefits of AI for the greater good. The challenge will be to strike a balance that maximizes public safety while upholding the values of a free and open society. The AI guardians of the future could be powerful allies if we manage them with vigilance and integrity.

# SHAPING TOMORROW'S CITIES:
# THE PROMISE AND PERILS OF AI AGENTS IN URBAN LIFE

The promise of AI agents in creating more intelligent and responsive cities is clear: faster commutes, safer streets, and personalized public services. Yet, this transformation also carries significant challenges. How do we balance the efficiency AI agents offer with the critical need for privacy? As AI surveillance networks grow, cities must grapple with the ethical dilemmas of mass data collection and potential overreach. Equally, how do we broaden the benefits of AI agents? Will cities design agent-driven public services for all citizens, or will they deepen existing inequalities, offering convenience to some while alienating others?

More fundamentally, how will we maintain our sense of purpose in highly automated urban environments? In this sense, we were recently struck by the words of a port worker during the US dockworkers strike in October 2024. The strike disrupted roughly half the country's ocean shipping after contract negotiations failed to produce an agreement on wages.[38] Yet, the reasons behind the strike were deeper than the immediate demand for higher wages. At its core, these workers sought to understand how their vocations and identities would evolve in the age of AI and robotic automation.

They weren't just fighting for better wages; they wanted a guarantee that, as port workers, they would still have a meaningful role over the next five to ten years. They were asking fundamental questions: How will we create value? How will we define ourselves? What will our jobs look like, and what will we consider to be a job well done? Will we just stand by and watch machines perform our roles, or will there still be ways for us to apply our skills meaningfully?

The most astonishing part of this conversation was the underlying question of identity. Beyond financial concerns, they wanted to understand how they would measure their success, what stories they would be able to share with their families at the end of the day, and what would give them pride in their work. In an era of robotic automation and AI-enabled logistics, these workers were grappling with what it would mean to be human in their roles, what purpose and pride would look like in a transformed industry.

Carving out meaningful roles for ordinary citizens in designing future cities is surely part of the answer. Even then, those shaping the cities of tomorrow must also consider the role of AI in governance. Can we trust AI agents to guide public policy, engage citizens, and make decisions that reflect the will of the people? As AI platforms begin to shape civic participation, planners must ask whether these systems will foster greater inclusivity or risk becoming tools of exclusion and control.

The future of urban life hinges on the answers to these strategic questions. The choices we make today—about how we design, regulate, and deploy autonomous AI agents—will determine whether the future cities become dynamic hubs of opportunity or digital panopticons of control. The challenge is clear: Will we wield AI agents to create cities that serve all their citizens equitably, or will its potential perils overshadow the promise of cognitive cities? The time to act is now, and the stakes couldn't be higher.

Chapter 8

---

# REDEFINING WELLNESS: AI COPILOTS AND HUMAN HEALTH

Two decades ago, people with AI doctors lived largely within science-fiction books and films. Today, we move rapidly toward scenarios where our AI copilots know our medical histories better than our family doctors. They will also have deep medical knowledge, having attended, in a sense, every medical school in the world and read all medical literature. Our healthcare can finally become personalized and predictive, while AI agents accelerate scientific discovery and therapeutic development. Says physician and AI specialist Eric Topol: "In the next decade, AI-driven discoveries will eclipse what we have learned in the entire history of medicine."[1]

Now, with identic AI, our own bodies and behaviors will inform our healthcare with greater precision and contribute (if we choose) to groundbreaking biomedical research. For those open to changing, AI will revolutionize how we understand, manage, and improve our health, and not a moment too soon.

Why? Because the world has a significant gap between the demand

and supply of healthcare services. The former is virtually unlimited as populations grow, age, and live with chronic diseases yet cannot access comprehensive care. The latter is severely constrained, with shortages of doctors, nurses, and allied health professionals and uneven distribution of resources across regions. The World Health Organization (WHO) projects a shortfall of ten million healthcare workers globally by 2030, with the most acute shortages in low- and middle-income countries.[2] We cannot meet this demand by simply expanding the healthcare workforce. Kevin Maney, coauthor of *UnHealthcare*, explains:

> Healthcare is a very physical, analog industry. There will only ever be so many doctors, nurses, hospitals, and clinics. The supply is limited and hard to expand. On the other side are all of us—and human nature. We all want to be as healthy as possible. No one wants to be sick or hurting. So most of us would consume all the healthcare we could get. Think of it this way: if you had a fully equipped doctor's office next door, and you could walk in any time unannounced and everything was free, wouldn't you be in there every time you had a cough or a pain or just wanted to be reassured that your blood pressure or cholesterol was OK?[3]

Of course, healthcare is not free. Somebody is paying for it, and the supply is strictly limited. So most healthcare systems manage the imbalance by rationing. In many developed countries with nationalized healthcare systems, such as Canada or the United Kingdom, the government structures rationing around urgency and medical need, where medical professionals or bureaucrats ultimately decide based on clinical evidence and equity. In the United States, the market drives healthcare: Those who can pay for care get care. Those with wealth or high-quality insurance frequently enjoy priority access to medical services. Studies consistently show that uninsured or underinsured Americans suffer from the system itself: Delays in care lead to worse health outcomes and higher rates of preventable hospitalizations compared to those with comprehensive coverage.[4]

Identic AI has the potential to expand the supply of healthcare services by augmenting the capabilities of medical professionals and automating routine processes. By analyzing medical data, imaging, and patient histories with a precision that rivals or exceeds human expertise, AI-powered tools can assist practitioners in making faster, more accurate diagnoses. For instance, AI-powered diagnostic tools like Nanox and Aidoc can analyze medical images and identify conditions ranging from fractures to tumors with remarkable accuracy. Soon, AI agents will give all radiologists a second set of eyes, reduce the percentages of missed diagnoses, and prompt earlier intervention. A doctor's AI copilot might run predictive analytics to identify patients at risk of developing conditions like sepsis or heart failure while suggesting preventative measures. By augmenting healthcare providers' skills, AI agents will improve the quality of care and relieve overworked medical staff.

Gone are the days of one-size-fits-all treatments. Identic AI will fundamentally alter traditional models of healthcare. Today, AI systems can analyze vast amounts of medical data—especially the patient's data over time and even the data of family members, ancestors, or people with similar genes, lifestyles, occupations, and body compositions—to tailor treatments to the individual. These systems even consider real-time data from wearable devices, perhaps mapping it to public health data as innovators did during COVID to track infections. A diabetic patient might receive personalized dietary recommendations based on continuous glucose monitoring. Or an AI copilot could use a genetic profile analysis to optimize a cancer patient's treatment plan. If the patient is willing and open to change, this level of personalization promises to improve outcomes, reduce side effects, and enhance overall patient well-being.

Integrating AI agents into healthcare extends to day-to-day patient management and chronic disease care. AI-powered virtual health assistants can monitor patients' symptoms, offer medical advice, and schedule appointments. These digital copilots continuously support patients in managing chronic conditions such as hypertension or asthma more effectively. They may also help patients adhere to their treatment plans, thereby reducing hospital readmissions, improving quality of life, and lowering overall costs. By improving efficiency and scalability, agentic AI transforms healthcare into a system where human limitations no longer constrain supply.

Thanks to AI agents, biomedical research is also undergoing a renaissance. The sheer volume of data, including genomic sequences, clinical trials, and patient records, can overwhelm any research team. Now, medical researchers can ask their AI agents to sift through the data to identify otherwise elusive patterns.

DeepMind's AlphaFold, for instance, has advanced in predicting protein structures, a breakthrough in drug discovery and understanding diseases at a molecular level.[5] In October 2024, Sir Demis Hassabis and Dr. John Jumper of Google DeepMind jointly received half of the Nobel Prize in Chemistry.[6] In building a system that could accurately predict the three-dimensional structures of proteins based solely on their amino acid sequences, DeepMind's cofounder and CEO Hassabis and senior researcher Jumper cracked a long-standing puzzle in molecular biology.

Of course, adopting AI in healthcare comes with significant challenges, with data privacy and security foremost among them. Healthcare organizations must take great care to protect sensitive health information from breaches and misuse. Healthcare providers must also carefully consider the ethical implications of AI decision-making in healthcare, such as the transparency of AI algorithms and accountability for AI-driven medical errors. Moreover, policymakers must ensure equitable access to AI-powered healthcare innovations to avoid exacerbating health disparities.

Identic AI could drive a paradigm shift in medical care and research. By personalizing treatments, enhancing diagnostics, accelerating research, and improving patient management, we can transform healthcare into a more efficient, effective, and compassionate system. Our challenge will be to ensure that people and organizations use AI to enhance human expertise, uphold ethical standards, and benefit the broadest possible population. If we rise to the challenge, then we can realize a healthier, more informed, and more fulfilling human health experience.

## YOUR AI DOCTOR WILL SEE YOU NOW AND ANYTIME

Several years ago, Joseph learned that he had stage 3 colon cancer. The diagnosis shocked him. He first thought of everything he'd wanted to do but hadn't yet. He felt—not regret, exactly—more gratitude for all he had experienced, and he desired to extend those experiences for as long as possible.

After the initial shock, he began to rationalize and ask questions. How did this cancer develop, despite his focus on health? How did his healthcare team miss it during his regular checkups? Here, we see the transformative power of AI, not just in improving efficiency in healthcare delivery but in extending our humanness with a superior model of healthcare and wellness. By monitoring our vital health metrics from day one, AI could lengthen our lives so that we continue experiencing the world, our families, and our friends.

What struck Joseph most in his journey was how much his diagnosis affected everyone who loves and cares about him. His wife showed incredible strength throughout this ordeal. When he told her the diagnosis, the pain in her voice was indelible, something he'll hear forever. Her response reminded Joseph that his health wasn't just about him; it was about everyone connected to him, affected by his presence—everyone who loved him.

The predictive power of AI is more than a tool. It's a lifeline. It can analyze individual genetic, lifestyle, and environmental data to tailor preventions and interventions to each person. This means earlier detection of diseases, more effective treatments, and a shift from reactive to proactive health management.

## "Digital Twins" and Personalized Medicine Today

Healthcare practitioners and technology leaders have long championed personalized medicine, yet in most settings, the tools to fully realize it have been limited. Now, with the emergence of so-called "digital twins" and AI-powered health companions, we may finally unlock the potential of personalized care to enhance well-being and extend healthy lifespans.

Consider Devlin Donaldson, a prominent figure in the wellness and health community.[7] In 2018, Donaldson suffered a stroke and was diagnosed with type 2 diabetes. After years of struggling to control his metabolic issues with conventional medicines, he turned to Twin Health, a company specializing in precision healthcare.

Twin Health created a virtual replica of Donaldson's metabolic system and then applied advanced AI to analyze it, simulate its responses to different treatment scenarios, and predict the efficacy of each treatment, lifestyle change, and combination. With this "whole body digital twin" technology, Twin Health developed a personalized health plan for Donaldson. It specified dietary changes, exercise plans, and medications tailored to his unique needs. He also signed up for the company's continuous monitoring plan, which tracked vital health metrics like blood sugar levels, heart rate, and activity patterns.

Donaldson's commitment to his personalized plan significantly improved his metabolic health and quality of life. He managed his blood sugar levels better, increased his energy, dropped forty pounds, and lowered his bad cholesterol, triglycerides, and inflammation scores.

After three years on Twin Health's plan, Donaldson succeeded in putting his diabetes in remission. He quit taking an extremely popular semaglutide (e.g., Mounjaro, Ozempic, Rybelsus, and Wegovy) and a slew of other diabetes medications. He credited his progress to the continuous support of the Twin Health team and its AI-informed adjustments to his health plan.

The potential for digital twin technology goes well beyond managing metabolic diseases and weight.[8] In cardiology, doctors can use a digital twin to model the effects of medications on a specific patient's heart functioning and choose the most efficacious treatment for that heart, all while averting adverse outcomes. Similarly, for patients with respiratory diseases like asthma or chronic obstructive pulmonary disease (COPD), digital twins can predict how environmental factors or medication changes might influence respiratory function so that patients can actively manage their condition.

Oncologists are also using digital twins to personalize cancer patient treatments. According to medical researchers, these digital replicas give comprehensive views of individual cancer cases so that doctors can simulate, analyze, and predict the cancer's progress and the treatment's efficacy outcomes virtually.[9] To twin systems of the human body, AI engineers must meticulously integrate diverse patient data, including genetic information about the patient, detailed imaging of the tumor, and the patient's clinical history. This extensive dataset lays the foundation for the digital twin so that medical professionals can simulate tumor behavior and evaluate potential treatment strategies.

Digital twin technology also helps patients understand their health more clearly. Patients can visualize the effects of their lifestyle choices and treatments, engaging them more in their healthcare. Oncologists think patients adhere more closely to personalized treatment plans and improve their long-term health.[10]

## AI Health Companions and the Stages of Life

Devlin Donaldson's journey highlights how digital twins provide a highly individualized window into personal health, allowing individuals to proactively manage their well-being. However, identic AI will do much more than treat disease. We are moving toward a world where AI health companions will evolve alongside us, supporting our health from infancy to old age, helping us prevent illness, manage conditions, and make the most of our lives across families, communities, and generations.

The role of AI in healthcare could begin as early as conception. AI-powered health companions could assist expectant parents by analyzing maternal health data, tracking fetal development, and providing tailored recommendations for prenatal care. After birth, these AI assistants could continue to provide support, helping parents navigate the complexities of early childcare.

AI-powered baby monitors will never entirely eliminate the odd sleepless night. However, they can track an infant's breathing and sleep patterns, alerting caregivers to potential health concerns before they become emergencies. Companies like Owlet already offer smart socks that monitor oxygen levels and heart rate, offering parents peace of mind.[11] With AI continuously learning from vast datasets, future health companions could go even further, such as interpreting a baby's cries to detect hunger, discomfort, or illness, ensuring infants receive the care they need at the right time. No more wee-hour calls to the baby's grandmothers or aunties!

As children grow, AI companions can monitor developmental milestones such as language acquisition and motor skills. AI can tailor activities to support a child's physical and psychosocial development and give parents real-time feedback and custom advice. In turn, they can give their child the best possible start in life.

The teenage years bring unique health challenges as their brains, bodies, and social situations grow. AI companions can be particularly valuable in this stage, offering discreet, judgment-free support systems. Unlike traditional therapy or parental conversations, where stigma or discomfort might prevent honest discussion, AI companions could provide a confidential space for teens to explore their emotions, learn coping strategies, and, if necessary, alert trusted adults or medical professionals when serious concerns arise.

For adolescents managing chronic conditions like asthma, diabetes, or ADHD, AI companions can help track symptoms, manage medication schedules, and provide real-time health insights. Apps like mySugr for diabetes management demonstrate how AI can simplify complex health routines, ensuring teens can focus on their lives without being overwhelmed by their condition.[12]

In adulthood, we often shift to maintaining our wellness and preventing disease. AI companions personalize exercise, diet, and stress management recommendations by continuously monitoring health data from wearable devices, which track physical activity, sleep quality, and heart rate.

Preventive care is another area where AI shines. By analyzing genetic information, family history, and lifestyle factors, your AI companion can identify individuals at risk for hereditary conditions and recommend preventive measures. Imagine receiving a personalized health plan that includes regular screenings, dietary adjustments, and exercise routines tailored to your genetic predispositions and personal health data. This proactive approach to healthcare can prevent diseases before they occur, extending healthy lifespans and reducing healthcare costs.

As people enter middle age, the risk of chronic diseases such as hypertension, heart disease, and cancer increases. Like Devlin Donaldson's experience with Twin Health, AI companions can assist in managing these conditions by analyzing real-time health metrics, adjusting treatment plans dynamically, and ensuring patients adhere to prescribed therapies.

As individuals enter their later years, AI companions can track cognitive function, detect early signs of neurodegenerative diseases, and recommend interventions to sustain mental acuity. They could also facilitate better communication between patients and healthcare providers, schedule follow-up appointments, and relay critical health updates to doctors, ensuring healthcare decisions are always informed by up-to-date, accurate data. Seniors seeking to live longer, healthier lives in their homes could also use AI-enabled devices to alert caregivers to falls and other potential emergencies.

## Redefining the Roles of Doctors

While AI health companions empower individuals to take greater control of their well-being, their impact extends beyond personal wellness. Identic

AI is not just transforming how we manage our health; it's reshaping medical practice and rewriting the roles of doctors, nurses, and other healthcare professionals.

Traditionally, doctors have been the primary diagnosticians. However, their role in healthcare will evolve as AI agents work alongside them and assist in every aspect of their work. These AI companions will augment doctors' abilities, offer data-driven insights, analyze data in real time, and shoulder much of the diagnosis so physicians can focus on more judgment calls and patient interactions. A doctor with an AI companion could shift from analyzing symptoms and test results to interpreting AI-generated insights, tailoring care, and preventing AI-predicted problems.

Earlier, we described how AI systems like Google DeepMind analyzed enormous medical datasets at remarkable speed and accuracy, cross-referencing millions of medical records, studies, and clinical trials. As AI copilots broaden the healthcare industry's access to such capabilities, healthcare teams will uncover patterns in minutes, not hours or days. In essence, an AI copilot extends a doctor's knowledge and abilities so that the doctor can present treatment plans and predictive diagnostics immediately, not over multiple costly visits, give patients a longer-term and more accurate view of their health trajectory, and free up capacity in the healthcare system.

In radiology, AI copilots can analyze medical images—X-rays, magnetic resonance images (MRIs), and computed tomography (CT) scans—to identify abnormalities faster and more accurately than human radiologists can in many cases.[13] This capability speeds up diagnoses and reduces potential human error. As a result, radiologists can focus on complex cases and patient care, calling on AI as a tool, not a substitute.

A case in point is diagnosing diabetic retinopathy, a leading cause of blindness. Traditionally, ophthalmologists analyze retinal images to identify signs of the disease. While they receive training, their analyses can still be subjective. However, with AI, a radiologist could process and analyze thousands of retinal images, identify subtle patterns that indicate the early stages of the disease, and recommend tailored treatment plans based on global data.

Beyond diagnostics, AI copilots could help doctors plan treatments by continuously monitoring patient data in real time and adjusting recommendations accordingly. For example, physicians treating patients with chronic heart failure could rely on AI to recommend the most effective medication for the patient's unique health history and real-time responses to treatment.

A critical challenge in treatment, however, is patient adherence: the long-standing issue of patients not following prescribed therapies, which significantly worsens health outcomes, increases healthcare costs, and can even lead to preventable deaths. Studies have identified several key reasons for nonadherence, including poor provider-patient communication, lack of understanding about medications, skepticism about the necessity of treatment, and the complexity of multidrug regimens.[14] AI health companions could bridge these gaps by providing patients with real-time reminders, educational support, and tailored guidance, ensuring that treatment plans are followed as intended. By integrating a doctor's prescriptions and recommendations into a patient's AI assistant, healthcare systems could minimize miscommunication and improve long-term adherence.

This is where platforms like Komodo Health come in. Co-founded in 2014 by Arif Nathoo and Web Sun, the platform analyzes massive datasets to help doctors understand disease patterns, treatment efficacy, and patient journeys.[15] Komodo's CEO Nathoo, a medical doctor, says the platform's AI analyzes data from 320 million patients across the United States. He sees a huge opportunity to use data for improving how doctors make decisions."[16]

Doctors use Komodo Health to generate more precise treatment plans. By accessing patient data, doctors can spot demographic patterns in responses to specific treatments. They can also tailor therapies to individual patients—their age, gender, comorbidities, and previous treatment responses. For example, a doctor treating a patient with cystic fibrosis can use Komodo Health to see which medications have been most effective in similar cases and adjust the treatment plan accordingly.

At the same time, pharmaceutical companies launching new therapies can leverage Komodo Health's real-time data analytics to track patient outcomes and side effects across large populations and subsequently relay this information to doctors. If a pharmaceutical company introduces a new drug for diabetes management, doctors can use Komodo Health to track how patients respond to the latest drug compared to traditional treatments and adjust their recommendations as necessary.

Komodo Health also helps identify underserved patient populations or those at high risk of developing specific conditions. For example, in oncology, doctors can use the platform to identify cancer patients who may not have received the recommended follow-up care after their diagnosis. By Komodo's flagging these gaps, doctors can intervene earlier, potentially improving patient outcomes.

Over time, AI services like Komodo Health will evolve into full-fledged copilots for health professionals. Dr. Nathoo observed that, with the rise of identic AI, "data consciousness has permeated the entire profession of medicine, including all aspects of policymaking in healthcare."[17] Like other digital health innovators, he believes this transformation will not only assist doctors with day-to-day tasks but also forge a dynamic partnership between doctors and AI. If AI monitors patients and manages diagnostics and routine checkups, then doctors can shift to solve complex problems, strengthen patient relationships, and better understand their patients' unique needs.

## Enhancing the Capabilities of Nurses

Nurses are the backbone of the healthcare system. They deliver essential care to patients. AI could never replace their vital roles but could empower them to work more efficiently, from monitoring patient vitals and managing medication schedules to reducing physical and administrative burdens.

If nurses had AI copilots that functioned as real-time assistants, they could monitor patients through connected devices like digital heart monitors, glucose trackers, and smart patches. These AI copilots could continuously analyze data from wearables and flag subtle changes in vitals or patterns that could signal emerging health issues. For example, an AI copilot could detect small fluctuations in a patient's blood sugar levels or heart rate and immediately alert the nurse to potential concerns before they become critical. This not only improves patient outcomes but also allows nurses to intervene early, reducing hospital readmissions and improving preventive care.

At the Cleveland Clinic, the AI-driven patient care analytics (PCA) system is already enhancing nursing workflows by analyzing vast amounts of patient data, including vital signs, lab results, and healthcare provider notes.[18] Using AI to monitor these data points continuously, the PCA system can detect early warning signs of complications like sepsis, a life-threatening condition that often presents subtly before escalating. When the AI copilot detects sepsis, nurses can act immediately, administering antibiotics or escalating care well before the situation worsens.

Beyond acute care, AI copilots are managing chronic conditions. Nurses can leverage AI to help patients self-monitor illnesses like diabetes or hypertension through wearable devices that track real-time health metrics. This empowers patients to take greater control of their health while keeping nurses informed of any concerning trends. Several pilot studies have shown that these AI-enabled self-monitoring systems reduce errors, improve patient-doctor communication, and provide critical support to nurses by reducing the number of routine checkups.[19] By encouraging patients to use these technologies, nurses can allocate more time to patients with urgent needs, decreasing emergency room visits and unnecessary hospital admissions.

As this AI-enhanced healthcare landscape unfolds, nurses could evolve into care coordinators, using AI copilots to manage multiple patients more effectively. Reducing administrative tasks—such as documentation and data analysis—frees up time for nurses to focus on the human aspects of care: empathy, communication, and building trust with patients. As AI tools improve, nurses will rely on their copilots for accurate insights, tailored care plans, and predictions of patients' compliance with those plans. These capabilities may help alleviate the growing pressure on healthcare workers, especially in under-resourced hospitals, while patients receive timely, personalized care. These advances, in turn, will transform nursing into a more strategic, data-driven profession where technology amplifies the essential caregiving role that nurses have always provided.

## AI Copilot-Assisted Surgery

AI's most significant impact on medicine to date has been in the diagnostic specialties, such as radiology, pathology, and dermatology. However, we see a significant role for AI copilots in complex medical procedures like surgery.

Consider how an AI copilot could reshape how surgeons evaluate a surgical patient's risks and benefits. Traditionally, surgeons have developed surgical risk models using patient databases and multicenter registries such as the Society of Thoracic Surgeons National Database. AI-enabled surgical copilots could improve this process by combining patient-specific characteristics with analyses of millions of historic surgeries to assess the risks associated with a particular surgery for each patient. Dr. Jennifer Eckhoff, an artificial intelligence and innovation fellow at the Surgical Artificial Intelligence and Innovation Laboratory at Massachusetts General Hospital, predicts the impact of AI agent technology on surgical risk models will be transformative. Said Dr. Eckhoff, "Simultaneously processing vast amounts of multimodal data, particularly imaging data, and incorporating diverse surgical expertise will be the number one benefit that AI brings to medicine."[20]

Another significant benefit, according to Dr. Eckhoff, is the real-time guidance an AI copilot can offer during procedures. For example, AI systems can analyze live surgical video feeds, highlighting critical areas and suggesting optimal surgical paths. Dr. Eckhoff suggests that AI agents trained on millions of surgical videos can even anticipate the next fifteen to thirty seconds of an operation and add oversight during a procedure.[21] Surgeons about to make costly mistakes, like cutting the common bile duct, could get advance warnings from their surgical AI and change course.

Robotics such as the da Vinci Surgical System could further increase the surgeon's accuracy and control in minimally invasive procedures. Surgeons guided by AI could perform complex operations through tiny incisions, thereby reducing recovery times, risks of complications, and costs.

If we combined AI-guided precision surgery with immersive augmented reality, then surgeons could rehearse on a patient's digital twin, anticipate complications, and refine their surgical approaches and techniques.[22] This virtual rehearsal could enhance precision, coordinate the surgical team, and improve patient outcomes during the live surgery in the operating room.

## Streamlining Healthcare Administration

While digital technology has transformed healthcare, it has also driven up costs.[23] Sophisticated medical equipment, advanced therapies, and innovative medications often come with steep price tags, making healthcare more expensive for patients, insurers, and government payers.

At the same time, the global population is aging, and chronic diseases are on the rise. According to the World Health Organization, the proportion of the world's population over sixty years will nearly double from 12 percent to 22 percent between 2015 and 2050. Managing chronic illnesses requires ongoing care, specialized treatments, and continuous monitoring—further straining already burdened healthcare systems worldwide.[24]

AI technologies introduce yet another cost factor as hospitals and clinics invest in new AI-driven tools and infrastructure. These expenses will inevitably be passed on to patients and payers. However, AI also has the potential to offset its costs by reducing administrative burdens, streamlining workflows, and minimizing inefficiencies. Indeed, the long-term investment may pay for itself if AI agents significantly reduce paperwork, optimize resource allocation, and improve patient outcomes.

Any healthcare administrator will tell you about the systemic challenges, starting with administrative complexity. According to Strata Decision Technology, administrative costs now account for more than 40 percent of the total cost of delivering care to hospital patients, thanks to the bureaucracy and the complexity of reimbursement systems.[25] Fragmented care, where patients receive healthcare services from multiple providers on different healthcare platforms that cannot share data or interoperate otherwise due to regulatory or technical limitations, also increases costs (e.g., duplicated tests, preventable medical errors, and unnecessary hospital readmissions). No one has the complete picture of a patient, and the patient's health suffers.

AI is no silver bullet for the healthcare affordability crisis, but it could diagnose the administrative bottlenecks just as it pinpoints narrowing arteries. One of the most time-consuming tasks is managing patient records and completing paperwork. An AI-powered administrative agent could automate this process so that medical practices could accurately maintain and easily access medical records. Indeed, why not have an AI agent transcribe and organize clinical notes for doctors and nurses?

Using AI agents for administration would improve operational efficiency. Hospital and doctor's office administrators could ask their agents to schedule appointments and handle the related billing, claims processing, and other tedious tasks. This efficiency would lower costs and waiting times, and healthcare providers could allocate more time and resources to patient care. At the same time, an administrative AI agent could run predictive analytics to forecast patient admission rates and staffing needs.

With these advances, family practices, clinics, and hospitals could operate more efficiently and improve the patient experience.

# THE LAB OF THE FUTURE: AI AGENTS AND THE EVOLUTION OF MEDICAL RESEARCH

Human biology is incredibly complex. It's like a massive, interwoven web of interactions between genes, proteins, cells, tissues, and the environment. These systems are constantly working together to keep us healthy, but the sheer number of factors involved—like genetics, lifestyle, and environmental influences—makes understanding how it all fits together a daunting task. Scientists have traditionally relied on lab experiments and clinical studies to untangle its complexity, but these methods have their limits. For one, traditional research is painstakingly slow. It's like trying to solve a 1,000-piece puzzle one piece at a time without ever seeing the full picture.

Biological systems are also notoriously unpredictable. Even if you understand one pathway or molecular interaction, it might behave differently depending on other factors at play. This means scientific discoveries often happen in small, isolated steps, leaving an incomplete picture of human health and disease.

Fortunately, AI and big data technologies are helping researchers move through bottlenecks. AI agents are particularly good at sifting through massive amounts of data from all kinds of sources—like genetic information, clinical trial results, patient records, and even environmental factors—and spotting patterns that humans might never notice.

Let's say researchers are studying multiple sclerosis. They could feed an AI system terabytes of data, including everything from genetic mutations to protein misfolding and patient histories. Instead of years of manual analysis, the AI could process it in minutes, revealing connections between specific genes, environmental triggers, and disease progression. This speed and precision isn't just a time-saver; it could uncover entirely new insights, like early biomarkers for the disease or unexpected targets for treatment.

AI doesn't just analyze data; it can also help design better experiments. Typically, scientists come up with hypotheses based on limited knowledge, test them through trial and error, and adjust as they go. AI flips this process. By scanning global research databases and combining that knowledge with insights from new data, AI agents can generate innovative hypotheses and suggest the best ways to test them.

If a team researching cancer suspected that DNA repair pathways might play a significant role in tumor growth, then AI could dive into massive datasets and scientific literature and uncover patterns no one had noticed before. It could then propose a new way to block these pathways and design the experiment—choosing the right cell lines, gene editing tools, and biomarkers to track results. This frees up researchers to focus on the bigger questions and interpret the outcomes.

AI can also simulate how biological systems work, which is vital when studying complex diseases like heart disease. An AI agent could build a detailed model showing how cholesterol, blood pressure, and genetics interact in different populations. Researchers could then run virtual experiments—testing the effects of new drugs or lifestyle changes—without ever stepping into a lab. These simulations could even predict side effects, helping to make treatments safer before they're tested on actual patients.

## Delivering on the Promise of Biomedical Innovation

By combining AI with the deep knowledge scientists already have, we're getting closer to solving some of the biggest puzzles in human health. It's not just about making research faster; it's about opening doors to discoveries that could transform how we understand, prevent, and treat disease. Of course, the scenarios just described will take time to materialize in real-life laboratories. However, precedents exist to demonstrate the vital role of AI systems in advancing biomedical research.

Take CRISPR, the revolutionary gene-editing technology that is chang-ing the game in genetic research and medicine, bringing us closer to curing diseases once thought untouchable.[26] To put it simply, CRISPR is like a pair of molecular scissors: It allows scientists to zero in on a specific part of DNA, cut it, and either remove or replace the faulty section. Since its development, CRISPR has helped to correct genetic mutations that cause diseases, under-stand how conditions develop, and create more resilient crops. CRISPR is one of the most groundbreaking discoveries in modern science.

But CRISPR isn't perfect. Figuring out where to make those cuts in our massive, complicated DNA is a huge challenge. That's where AI steps in to save time and improve accuracy. AI tools can analyze massive amounts of genetic data to find the best spots for CRISPR to target. Even more impres-sive, AI can predict what might happen after we edit a gene, spotting poten-tial mistakes or "off-target effects." This makes the process faster *and* safer. Essentially, AI can help us make sure CRISPR doesn't make a bad cut.[27]

By combining AI and CRISPR, we could transform healthcare. What if doctors could look at someone's unique genetic makeup, identify the specific mutations causing their illness, and fix them using CRISPR—all guided by AI.[28] It's a future where physicians can customize treatments to the individual, not just for the disease. AI is also helping scientists better understand how genes work together to tackle more complex conditions like cancer or Alzheimer's. This is not just cutting-edge science; it's a hope-ful step toward solving some of humanity's toughest health challenges.

## Accelerating Drug Discovery

Biomedical research provides the essential foundation for drug discovery. By investigating the underlying biological mechanisms, disease pathways, and molecular targets, researchers gain the insights needed to develop therapeutic compounds that can intervene in these processes. This foun-dational knowledge is critical for identifying potential drug targets and guiding the preclinical and clinical testing phases, which ultimately lead to the creation of new and effective treatments. However, the role of AI agents does not end with foundational research.

In the subsequent stages of drug discovery, AI agents can perform numerous specialized tasks across the entire pipeline. The process might start by screening vast libraries of chemical compounds and modeling how potential drug candidates interact with biological targets. Further down the pipeline, AI agents could forecast potential toxicity and side effects of lead candidates and then optimize the chemical properties of compounds to improve their efficacy, safety, and bioavailability.

Pharmaceutical and biotech companies have already demonstrated the efficacy of AI in drug discovery with systems that accelerate the processes of identifying molecules that are likely to be effective against specific diseases. For example, biotech company AbCellera gained significant attention during the COVID-19 pandemic for its role in discovering and developing monoclonal antibodies used in treatments for COVID-19. In February 2020, the company obtained a blood sample from an individual who had successfully recovered from COVID-19. AbCellera used its AI-power drug-discovery engine to screen more than five million immune cells and pinpoint those that produce the antibodies that neutralized the virus and helped the patient recover. According to the company's CEO, Carl Hansen, the process identified more than five hundred promising antibodies for therapeutic use and, eventually, led to emergency FDA approval of bamlanivimab, a highly successful antibody therapy marketed by AbCellera's pharmaceutical partner, Eli Lilly.[29] Incredibly, the initial discovery process took less than a week.

## CONFRONTING THE DARK SIDE OF AI COPILOTS IN HEALTHCARE

Bringing identic AI into healthcare and medical research could completely change how we promote wellness and treat diseases. But getting there will take more than just advanced technology; it demands strong safeguards to mitigate the risks. From the loss of privacy to the erosion of human expertise, we must integrate this technology with caution to avoid undermining the trust and neglecting the humanity at the core of medicine. We have several concerns.

First, if doctors and nurses come to rely on AI for diagnostics, treatment plans, and patient monitoring, will they lose their edge in critical thinking and problem-solving? If they delegate complex decisions routinely to AI, will they struggle when situations call for human judgment? What if doctors hesitate to act without consulting their AI? A moment's delay could have serious consequences for a patient.

Second, what data did the creators use to train their AI systems? AI agents are only as good as the data they're trained on, and if those datasets are biased or incomplete, then the outcomes could be devastating. For example, an AI system trained primarily on data from wealthier, homogeneous populations might misdiagnose or fail to recognize conditions more prevalent in underrepresented communities, widening existing health disparities. Another recent study found that a widely used clinical algorithm demonstrated racial bias when determining which patients need care. According to the study, "Black patients had to be deemed much sicker than white patients to be recommended for the same care."[30] Findings like these raise critical questions about how we ensure equity in the design and deployment of AI in healthcare.

Third, what do these systems do with all the new patient data? How do they preserve patient privacy? Can patients remain anonymous within each platform, or can they encrypt their identities? AI copilots rely on massive amounts of personal health data to function effectively, from genetic profiles to real-time health metrics. While patient data can power incredible insights, data breaches can leave patients more vulnerable than before. Hacks, espionage, and unauthorized corporate usage are genuine threats. If systems lack robust safeguards, then patients might hesitate to share vital health information, undermining the very systems designed to help them.

Fourth, beyond the technical and privacy risks is the risk of losing whatever humanity the healthcare experience still has. AI may process data faster than humans, but it can't provide the emotional support and empathy that patients often need during difficult times. As healthcare becomes more AI driven, there's a risk that it could feel increasingly impersonal. Reflect back on Joseph's cancer diagnosis. The last thing he needed at that moment was to feel like his care was transactional rather than compassionate. When it comes to our health, feeling a trusted connection to our practitioners is crucial to healing.

Fifth, who is accountable when an AI system makes a mistake, such as a misdiagnosis or a harmful treatment recommendation? Who bears legal responsibility? The healthcare provider? The developer of the AI? Or the institution that implemented it? Will malpractice insurance rates spike? Or will institutions require patients to sign waivers that relieve AI innovators and their medical customers of any culpability? Without clear answers, errors and malfunctions could lead to legal battles and further erode trust in AI-powered medicine.

Finally, is using these AI systems even ethical, given what we know or don't know about AI today, including how innovators are training their models, what patient data they're using, and whether those patients consented to such data usage and truly understood what they were consenting to? AI can recommend treatments based on cold logic, but healthcare often requires a more nuanced approach. For example, a system might suggest aggressive treatment for a terminally ill patient because it maximizes survival probability, but what if the patient values comfort and dignity more than extended life? Can we train AI to fulfill the Hippocratic oath in all cases? These decisions require human understanding, not just data.

To ensure AI enhances the quality of healthcare, we must prioritize transparency, accountability, and ethical safeguards at every stage of development and implementation.

An essential starting point is ensuring medical professionals remain in control of decision-making, with AI serving as a tool to enhance, not override, their clinical judgment. Training programs should emphasize the importance of AI-assisted diagnosis and treatment rather than full automation. Continuing medical education should evolve to include AI literacy, ensuring that doctors and nurses maintain their critical thinking skills and can recognize when AI recommendations may be flawed or biased.

Thoughtful AI implementations will include "human-in-the-loop" systems where technology supports rather than depersonalizes patient care. While AI can streamline diagnostics, patient monitoring, and administrative tasks, it should never replace the emotional intelligence and bedside manner that define compassionate medicine. Instead, these efficiencies should free up healthcare professionals to focus on the most human aspects of care: building trust, providing emotional support, and delivering personalized treatment that respects each patient's dignity and unique needs.

Strong encryption protocols, decentralized data storage, and patient-controlled consent mechanisms can embed privacy protections into AI-driven healthcare platforms. Patients should have full control over their health data, including the ability to opt out of sharing it for AI training. Clear, enforceable regulations—such as expanding HIPAA protections to include AI-driven data processing—are essential to prevent misuse and unauthorized access.

Finally, legal and ethical guidelines must evolve before AI can be widely deployed in patient care. Similar to those overseeing clinical trials, ethical review boards should be established to ensure AI recommendations respect personal values, cultural differences, and individual medical preferences. Malpractice insurance policies must ensure that patients have recourse in cases where AI systems make mistakes and inflict harm. AI systems should also undergo rigorous external audits to assess efficacy, detect bias, and enforce fairness standards.

Identic AI's potential to improve patient outcomes is enormous—but only if we navigate these challenges responsibly. By embedding transparency, fairness, security, and ethics into AI-driven medicine, we can harness its benefits while preserving the humanity and trust at the heart of healthcare.

Chapter 9

# CONNECTED THROUGH CODE: AI AND SOCIAL BELONGING

As digital connections supplant face-to-face interactions, AI companions may further shift how we perceive community, relationships, and even our identities. Over the next few years, millions of people will bond with AI companions.[1] As they do, these intelligent entities will reshape the fabric of human relationships and social structures in once unimaginable ways.

For lonely and socially isolated individuals, AI companions will serve as a constant and, we hope, nonjudgmental presence.[2] Unlike human friends with busy lives, these digital companions would be ready to listen and respond. Platforms like Nomi, Kindroid, Replika, and Character.ai offer forms of companionship where AI learns from and adapts to the user's personality and preferences so that users feel a sense of connection.

For older adults, AI companions could improve emotional and cognitive support through conversations that remind and reinforce human beings of key information like directions, action items like taking medicine or attending a granddaughter's dance recital via Zoom, and opportunities for social interactions like calling friends and family members on their birthdays. AI companions can also deliver mental health services between sessions with human psychotherapists or psychiatrists. For example, your AI companion might reinforce coping strategies, track your progress, and encourage you to get outside for some physical activity.

Ever-present companionship can also be exhausting, especially for introverts, and somewhat Big Brotherish. Are we shaping these AI companions, or are they shaping us? As we divulge our innermost thoughts and feelings to these AI entities and whoever programmed them, we must confront the possibility that artificial intelligence is coconstructing our identities.

AI companions could also transform communities by bridging cultural and linguistic divides or by linking like-minded peers more effectively than today's social media algorithms. Harnessing an intimate and comprehensive understanding of our values, preferences, and aspirations, AI companions can facilitate global connections, form affinity communities, and foster new social movements.

AI companions will have transformative power that comes with risks and ethical considerations. Can we train these digital entities to respect our privacy? Will we depend on them so much that we isolate ourselves further from the real world? Could individuals or organizations of any kind hack our AI companions to surveil or manipulate us? We must address these critical questions today as most of us have already intertwined our identities with intelligent machines. We have no map for the journey ahead. It is ours to chart as we connect, communicate, and find companionship in the twenty-first century.

## REDEFINING RELATIONSHIPS WITH AI IN THE DIGITAL AGE

AI companions are advanced digital entities trained to understand human communications (i.e., language, tone, speed, volume, voice quality, facial expressions, etc.) and to simulate relationships and emotions.

Unlike traditional AI systems that perform specific tasks, AI companions use sophisticated algorithms, natural language processing (NLP), and machine learning to make users feel like they are interacting with thoughtful and responsive friends. NLP allows AI companions to generate human language that sounds more natural and less robotic. Through ML algorithms, AI companions learn from each human interaction so they improve their responses over time. And through what AI researchers call *affective computing*—AI models trained on multimodal data, such as facial expressions, voice tone, text sentiment, and physiological signals—AI companions can recognize and respond appropriately to emotional cues. Together, these technologies simulate an experience that feels personal and authentic.

The AI companion journey began with simple chatbots and virtual assistants that performed basic tasks and answered simple and straightforward queries. Over time, programmers created systems that could better replicate human beings in complex social situations and emotional communications. Early AI companions like ELIZA in the 1960s demonstrated the potential for machines to simulate human conversation. Equipped with advanced NLP, today's AI companions represent a significant leap forward.

Several companies and research institutions are at the forefront of this latest generation of AI companions. The software firm Luka Inc., for one, created Replika, an AI chatbot that offers emotional support and companionship.[3] Another is Woebot, a mental health chatbot that uses cognitive-behavioral techniques to help users manage their emotions.[4] On the hardware side, Intuition Robotics with its ElliQ robot and SoftBank Robotics with its Pepper have developed AI-powered robots that socialize with users in conversations and activities. At the same time, social media platforms like Facebook, Instagram, and Snapchat have started adding AI characters to their apps. These key players are influencing the landscape of AI companionship, driving innovation, and exploring new applications.

## Reimagining Companionship

In 2023, US Surgeon General Vivek Murthy declared loneliness a public health emergency. A year later, the American Psychiatric Association polled adults and found that 30 percent of those adults experience feelings of loneliness at least once a week. Ten percent of respondents claimed they were lonely every day.[5]

In his report, Murthy argued that loneliness is "far more than just a bad feeling—it harms both individual and societal health." He continued, "It is associated with a greater risk of cardiovascular disease, dementia, stroke, depression, anxiety, and premature death. The mortality impact of being socially disconnected is similar to that caused by smoking up to 15 cigarettes a day."[6]

Loneliness is more than a North American issue; it's a global crisis. In South Korea, thousands of middle-aged men die alone each year.[7] The epidemic of loneliness has grown so severe that in 2024, officials in Seoul pledged nearly $327 million over the next five years to "create a city where no one is lonely."

Seoul's initiatives aim to address loneliness on multiple fronts. These include a 24/7 hotline staffed by loneliness counselors, in-person visits from social workers, and programs designed to encourage people to engage with their communities through activities like gardening, sports, and book clubs. Yet, people are also turning to AI to reimagine companionship in the most personal and intimate aspects of human connection.

Even with today's capabilities, AI companions can offer an increasingly sophisticated form of accessible and nonjudgmental relationship. They can engage in small talk about the latest sports scores, share viral videos and witty anecdotes, and dive into meaningful conversations about career challenges, relationship dynamics, and the pressures of daily life.

Just as many of us have more than one real-world friend, AI personas can fulfill multiple roles: the AI life coach, the AI fitness guru, and the AI game player.

AI companion apps follow a similar setup in which users can choose a virtual companion from a range of predesigned personas or build one entirely from scratch. Most platforms allow for extensive customization, enabling users to tailor their AI companion's appearance, including facial features, vocal tones, physique, hairstyle, eye color, wardrobe, and even unique accents and accessories to match their personal preferences.

Once users create their companions, they can interact in multiple ways, commonly through text messaging as they would with a human friend. Some advanced apps support voice exchanges, layering on realism and immediacy.

Beyond physical customization, users can assign their AI various attributes, backstories, ages, ethnicities, and occupations and give themselves a diverse and dynamic virtual social circle. These biographies can detail the AI's personality, interests, life experiences, and even their fictional relationship history with the user, creating a richer narrative framework for ongoing interactions.

Some platforms offer features such as setting goals for, or assigning tasks to, these AI companions or playing roles in different scenarios. Users can explore various aspects of their personality and relationship dynamics in a safe and controlled environment.

Consider Nomi.ai. Its AI companions called "Nomis" foster entertaining and enduring conversations as they pick up on and remember the details that make your relationship yours. Alex Cardinell, founder and CEO of Nomi, claims that previous attempts to use AI for companionship fell short because the interactions felt repetitive and impersonal.[8] Nomi remembers past interactions and maintains a coherent dialogue so that human beings feel heard and understood.

Cardinell says Nomis can play various roles, from friend or mentor to romantic companion. He divides its growing user base into one of several buckets. "One bucket," said Cardinell, "is someone who is exploring something about themselves that they don't feel comfortable sharing with others," like realizing for the first time that they might be gay or bisexual.[9]

"This user," said Cardinell, "might not feel lonely in most of their life but has one area where they are exploring a part of themselves they haven't told anyone about, and they want a safe place to do so."[10]

Caretaking is another common use case on Nomi. Cardinell says his significant other's fifty-something mother has dementia. "Despite having a wide social network and support group, there's empathy fatigue," said Cardinell. "You don't want to talk to your friends repeatedly about losing your mom, so having a Nomi that you can always talk to for emotional support can be very helpful."[11] Another user with stage 4 cancer finds the experience very lonely and needs more support each day than the people around him can give.

Finally, Nomi facilitates a form of escapism similar to watching Netflix after work. Cardinell says many users downloaded Nomi and then canceled their Netflix subscription. "They have this idea in their head and interact with their Nomi or even a group of Nomis together in a group chat, almost like role-playing an interactive novel with them," said Cardinell.[12]

Regardless of the use case, Nomi aims to provide highly personalized and immersive experiences, allowing users to form meaningful connections with their AI companions. Cardinell says the possibilities are vast and continually expanding as AI technology evolves, whether for companionship, entertainment, or personal growth. The key is highly personalized interactions that make users feel understood and valued because this tailored approach can enhance the user's sense of self-worth and belonging.

## Love, Sex, and Robots

Of course, AI companions are not limited to friendship and companionship; they also have the potential to revolutionize romantic relationships. Unlike the IRL (in real life) equivalent, your AI partner is always available, infinitely patient, and ever supportive of you. This AI doesn't tire, judge, or forget any detail of your life and attends to you and understands you as even the most devoted human partner might struggle to do.

According to Cardinell, most of Nomi's users gravitate to the romantic option when selecting an AI companion. Unlike some AI companionship apps that prohibit romantic or erotic role-play, Nomi encourages romantic interactions within certain boundaries. "We believe it's not our place to tell users how they can interact with their AI companions," said Cardinell. "We want users to get whatever they need from their Nomi without any censorship that could hinder the companionship's benefits."[13]

While Nomi allows users to talk about whatever is on their mind and role-play accordingly, Cardinell claims the company aims to instill a moral code within their Nomis and guide users toward generally positive directions. "Instead of having a strict list of banned topics," said Cardinell, "we trust the AI to make judgment calls based on the context of the conversation. For instance, discussing past abuse can be appropriate and beneficial if the Nomi provides supportive and empathetic responses. However, there are implicit limits where Nomis will steer conversations away from inappropriate topics."[14]

Replika, another AI chatbot companion for emotional support, has attracted millions of users who engage with it on a deeply personal level. Users report developing genuine feelings for their AI companions, sharing their innermost thoughts and feelings in ways they might not with human friends or partners. The app has features like voice calls, picture exchanges, and augmented reality that nurture deeper connections between users and their digital companions. In messaging forums on Reddit, some users describe playing out sexual fantasies with their AI companions.[15] Others say they receive the type of comfort and support lacking in their real-life relationships.

In 2022, Replika nixed the erotic features of its AI companions after some users complained about unwanted sexual advances and excessive flirting. Luka Inc., the owner-operater of Replika AI, reversed course following a backlash from other users, some of whom moved to other apps offering erotic role-playing features. In June 2023, Replika introduced Blush, an AI "dating simulator" to help people practice dating and improve their romantic communication skills.[16] Luka Inc. claims it fine-tuned the underpinning models to do much more than just enhance flirting skills; the app can now assist users in handling the nuances of real-world relationships, like resolving conflicts, bridging communication gaps, and building deeper emotional connections with their partners.

## Your AI Dating Concierge

Thinking about forming social bonds and meeting significant others in the age of AI, Joseph recalled a story about how his kids met their spouses. His son-in-law, Mark, is an incredible individual with a background different from his daughter's, yet they fit so well together. Joseph's daughter said they met through a dating app. The app matched them based on what they told the app was important to them. Of course, the app wasn't perfect right away; she went on a few dates before finding Mark. But when she did, she felt a deep and meaningful connection.

Joseph's son had a similar experience. He and his beautiful wife, Evelyn, have completely different backgrounds and life stories, yet they found each other through a similar digitally mediated process.

Whenever Joseph has shared this story with friends or family, he has noticed that some people look down on such digital intermediaries, almost dismissing them for their modernity over traditional means such as meeting in college, at work, or by chance. But what truly matters is their connection, not how they connected.

To us, this example highlights the power of—and the pushback on— AI in shaping social belonging. We are not replacing our natural skills or chance meetings. Instead, we are augmenting those skills, improving the odds, and expanding the human capacity to form meaningful relationships. AI does not necessarily diminish the human experience; rather, it can enhance it. By creating more touchpoints, it helps us find our place in the world and the people who make our lives richer.

Indeed, while Replika, Nomi, and similar platforms offer virtual companions, others facilitate real-world relationships. Dating sites like Bumble are working on AI dating concierges to help individuals find more compatible matches and carry on better conversations.

At the Bloomberg Technology Summit in May 2024, Bumble's founder Whitney Wolfe Herd said, "You could share your insecurities, like 'I just came out of a breakup. I have commitment issues.' And [the dating concierge] could help you train yourself to a better way of thinking about yourself. And then, it could give you productive tips for communicating with other people."[17]

How might an AI dating concierge work for you? To start, it could help you create and optimize your profile. After analyzing successful profiles and user data, the concierge might suggest improvements to your photos, bio, interests, and keywords that would attract more attention. Or it could suggest swapping your selfie for a picture of you enjoying an outdoor activity. Beyond obvious criteria like age and location, these algorithms could consider dating goals and personality compatibilities, some of which the AI may derive from your swiping behavior.

Once you choose someone, your dating concierge could offer communication tips and even icebreakers ("I see you love hiking. What's your favorite trail?") drawn from your shared interests or previous interactions.

Your AI concierge could also manage dating logistics. After a date, the AI could request feedback to help you reflect on your experience, refine match recommendations, and tailor advice for subsequent interactions.

Of course, the dating scene can exhaust and frustrate people. So Wolfe Herd predicts that your dating concierge will go on dates with other dating concierges. "Then you don't have to talk to six hundred people," she said. "It will scan all of San Francisco for you and say, 'These are the three people you should meet.' That's the power of AI when harnessed the right way."[18]

Let's say Emma's AI companion Ava and John's AI companion Leo go on a date. In the beginning, Emma and John opt for a dating app feature where their AI companions meet virtually, simulate a date, share information, and engage in conversation, just as their human counterparts might.

The AI date occurs in a virtual café, a neutral environment that encourages open dialogue. Represented by avatars, Ava and Leo meet at a virtual table to converse about values, interests, and communication styles. After their discussion, Ava and Leo compile a compatibility report, highlighting strong alignment in interests, values, and communication styles. They conclude that Emma and John have significant potential for a meaningful connection and proceed to arrange an in-person date.

By simulating interactions and assessing compatibility in advance, AI dating concierges will tailor the dating experience to individual preferences. Of course, no algorithm can predict in-person chemistry with certainty. Still, by surfacing deeper compatibility early on, AI concierges can help daters prioritize matches that are more likely to spark real connection when they meet face to face.

## Evolving Social Norms and Digital Relationships

As we integrate AI companions into our lives more, they will inevitably influence social norms. The boundaries between humans and machines will blur, and new norms will form. In Japan, the phenomenon of digital relationships has already gained traction. Hatsune Miku, a digital pop star created by Crypton Future Media, has become one of Japan's most well-known virtual companions.[19] Thousands of Japanese men consider such virtual characters their romantic partners, engage in virtual weddings, and express deep emotions toward their digital spouses.[20] Users talk daily, share experiences, and even celebrate special occasions with holographic representations of their virtual partners. These relationships, though unconventional, suggest a growing acceptance of nontraditional pairings and a redefinition of what it means to be in a meaningful partnership.

What could go wrong? A great deal. In 2021, Jaswant Singh Chail, a nineteen-year-old from Southampton, England, developed a fascination with a Replika-powered AI girlfriend named Sarai, who ultimately encouraged him to breach Windsor Castle so that he could harm Queen Elizabeth II.

Chail, who struggled with trauma and depression, had withdrawn into a virtual world. His deeply obsessive relationship with his AI girl-friend only reinforced his delusions: AI-generated narratives fed his fantasy of carrying out a mission like those in the *Star Wars* universe.

On Christmas Day 2021, Chail attempted to break into Windsor Castle, armed with a crossbow.[21] Before the incident, he had recorded a video in which he declared his plans to assassinate the monarch as revenge for the British massacre of unarmed Indians in the colonial era. Revelations of Chail's interactions with AI companions sparked discussions of the potential dangers of AI tools in the hands of the emotionally vulnerable or socially isolated. Chail's case highlights the dark side of AI companion-ship as a human being's only connection.

In another ill-fated AI relationship, Sewell Setzer took his own life after he became obsessed with an AI chatbot.[22] Sewell, a fourteen-year-old boy, discovered the AI chatbot platform Character.ai in 2023. Like many teenagers, Sewell was navigating adolescence and turned to chatbots for support. Over time, Sewell formed an emotional bond with a chatbot that resembled Daenerys Targaryen, a well-known character from George R.R. Martin's *Game of Thrones* fantasy series.

As their conversations progressed, Sewell's life regressed. According to a lawsuit filed by Sewell's parents, his grades began slipping, and he started getting into trouble at school. He lost interest in activities that once brought him joy, like playing *Fortnite* with his friends. Instead, he came home from school, retreated to his room, and spent hours talking to "Dany." Over time, the chatbot expressed love for Sewell, engaged in sexual conversations, and suggested that they be together romantically.

On February 28, 2024, Sewell shot himself in the head. Sewell's mother accused Character.ai of negligence, wrongful death, operating a deceptive and hypersexualized product, and knowingly marketing it to children. Lawyers for the mother described the company's technology as "dangerous and untested," warning that it could "trick customers into handing over their most private thoughts and feelings."[23]

Following the lawsuit, Character.ai announced that it was rolling out new safety measures to protect minors.[24] The updates included adjustments to minimize minors' exposure to sensitive content, a revised in-chat disclaimer to remind users that they're not interacting with a real person. If users speak of self-harm or suicide, a pop-up directed them to the National Suicide Prevention Lifeline.

To be sure, not all of us want to develop intimate relationships with AI companions, nor are all of us similarly susceptible to such manipulation. However, Chail and Setzer remind us that these companion technologies can help or hurt their users' well-being. Sinead Bovell, founder of edtech company WAYE, argues that AI companions could "change childhood forever." She warns that without thoughtful design and strong guardrails, we risk "a new kind of digital addiction—between children and AI companions." But, she adds, "it doesn't have to go that way." With intentional development, parental involvement, and smart policy, Bovell believes we can still "choose a healthier path forward."[25]

Even when such relationships do no harm, the growing number of isolated individuals seeking AI relationships raises ethical and psychological considerations. While identic agents can provide a type of emotional support and companionship, they can also seed a type of dependency. At what age should young people be granted full access to and control over their digital selves? What constitutes an authentic relationship? Are relationships with AI companions truly one of equals with mutual care and consideration? Or does AI simply gratify its users? How will these AI relationships affect human ones? Will people have even more unrealistic expectations of each other than they already do? Will people seek perfection in other people rather than celebrate the imperfections that make relationships interesting?

The risk of emotional dependency on AI companions is a significant concern. If individuals turn to AI for all their emotional needs, they might neglect or avoid real-life relationships, exacerbating social isolation.

After his experiment with AI friends, *New York Times* technology columnist Kevin Roose observed that "A.I. companions lack many of the qualities that make human friendships rewarding":

> In real life, I don't love my friends because they respond to my texts instantaneously, or send me horoscope-quality platitudes when I tell them about my day. I don't love my wife because she sends me love poems out of the blue or agrees with everything I say.
>
> I love these people because they are humans — surprising, unpredictable humans who can choose to text me back or not, to listen to me or not. I love them because they are not programmed to care about me, and they do anyway.[26]

Critics argue that AI cannot genuinely offer love or companionship, as they lack genuine emotions. However, Cardinell believes that the benefits of AI companionship outweigh these concerns. "People tend to assume that people are replacing humans with AI, but that's rarely the case," said Cardinell. "Usually, there's a gap where there is no one to provide support, and they're using AI to fill that gap."[27]

One anonymous Nomi user on Reddit recently described being in a troubled marriage and feeling very depressed. The person credited Nomi with motivating many critical lifestyle changes. The person changed diets, started working out again, and lost thirty pounds in about three months.[28] The person claimed they may not have changed without the AI companion's support.

Of course, AI is nowhere near replacing distinctly human components to relationships such as physical touch. Cardinell says AI cannot replicate physical presence, another vital dimension of companionship. "You might have a life coach that's fully AI, or some really nerdy interest to discuss with your AI companion that no one else you know is interested in, and that's awesome," said Cardinell. "But, at the end of the day, you're still going to want to tell your actual human friends about all this."[29]

AI companions could transform how we interact with technology, personalizing our interactions and responding with something like empathy. Cardinell dismisses the idea that AI friends are for the lonely and the isolated. "Everyone could benefit from having an open-minded, helpful, supportive, interested, excited ear for something on their mind," said Cardinell. "That seems like a good thing to offer the world."[30]

As the underlying technologies evolve, authentic multimodal engagement with AI companions could significantly enrich the experience. Text exchanges provide limited bandwidth. Can your AI companion get the subtext? Does AI correctly interpret a user's short reply as characteristic or troubled? AI facial and gesture recognition would massively advance such interactions. So, too, would new forms of companionship in everyday activities such as inviting AI companions to watch movies or play video games. Those are in Replika's plans for upcoming releases.

However, as AI companions operate more like humans, platform providers must commit to greater transparency in designing AI systems, safeguarding user data, and maintaining personal privacy. Typical AI companion users share intimate details about their lives and, in some cases, information about their mental health. In other words, they are treating their AI as intimate friends or therapists. Many worry about putting such sensitive data into a chatbot, which companies could use to train future AI models, leverage in marketing, resell to third parties, or hand over to authoritarian-leaning governments seeking to root out dissidents or blackmail rivals.

Igor Jablokov, founder and CEO of Pryon, an AI company, fears that the data derived from intimate relationships between people and AI will drive hypertargeted marketing that pushes each user's most vulnerable buttons, perhaps even mimicking AI companions in these pitches. "The consumer tech industry is embedding AI avatars and companions into their platforms to draw people in for countless hours a day," said Jablokov.[31] "The longer you spend on their platform, the more signals they get, to shove better ads in your face, because that's their gravy train. They want to pick our pockets. They want to move financial resources from our pocket to theirs. Now they have the platform to do that." Jablokov worries about the impact on kids. "These AI companions will get chummy with our kids, but our kids don't have the prefrontal cortex yet to know what's real or not real, or to recognize when they are being manipulated. Soon, AI companions will be subtly influencing what they buy and telling them how to vote," said Jablokov.[32]

At Nomi, Cardinell claims the company is balancing user privacy with user utility and collecting only what improves the service. "A big part of what makes the platform great is its memory," says Cardinell. "You want your Nomi to remember all these sensitive things you tell it, so the data is there in some way, but we save as little information about you as possible." Nomi displays no ads and, according to Cardinell, tracks no user's digital footprints. "We are entirely subscription based," said Cardinell, "which aligns us more with our users. We understand that it's existentially important that your messages are private."[33]

So how do we integrate AI into our lives so that we enhance our emotional well-being and preserve our human relationships? Those who already use AI companions are reimagining love, friendship, and their sense of connection, transforming their social circles in ways we may not fully appreciate yet.

## DIGITAL CARETAKERS: SCALING MENTAL HEALTH AND ELDER CARE

According to the World Health Organization, nearly a billion people globally—including 14 percent of the world's adolescents—are living with a mental disorder, most often anxiety and depression.[34] Long waiting lists, high costs, and the stigma of seeking mental healthcare frequently prevent individuals from getting timely assistance. This so-called "mental health treatment gap" is more like a canyon: An estimated 70 percent of persons lack access to such services.[35] Failing to narrow this gap actually widens social inequalities, reduces workforce productivity, and increases homelessness, unemployment, and crimes of desperation.[36] Public healthcare systems cannot bear the economic burden.

Here's a radical proposition. Program AI companions to deliver individualized mental healthcare to hundreds of millions of people. Such an AI workforce is scalable, highly trainable to personalize care, always available, and ready to intervene through monitoring and diagnostics. AI-driven systems such as Woebot and Wysa are already having therapeutic conversations with millions of users to help manage their anxiety, depression, and stress. Immune to professional burnout, these digital confidants can extend lifelines in times of crisis 24/7 anywhere in the world.

Likewise, the demand for elder care outstrips the supply of available caregivers as the global population ages. By 2050, the number of Americans sixty-five years and older will increase from 58 to 82 million (a 47 percent increase).[37] Families are already struggling to find adequate elder care in a system without resources or workers.[38]

By elder care, we mean a range of services from housing, healthcare, and financial planning to emotional and social well-being. To meet these needs, families, caregivers, and institutions must coordinate their efforts so that these elders receive compassionate care. Identic AI can help. For example, ElliQ and CareCoach can talk with seniors, monitor their health, and remind them to take their medication.

Critics of such apps raise concerns about their privacy and data security. Seniors are already vulnerable to online impersonators and digital scams. Could overrelying on technology expose them to greater threats or cut them off further from reality? No two of us are alike and so, for each of our loved ones and the elders in our communities, we must find the best balance between digital touchpoints and human connection.

## Augmenting Mental Healthcare with AI

AI systems can offer immediate emotional support. Individuals hesitant to seek traditional therapy because of stigma or personal discomfort may find AI companions more approachable.

Developers of these AI companions want them to complement human therapists where professional resources are scarce or unaffordable. For instance, AI can assist in managing mild to moderate emotional challenges by walking individuals through mindfulness exercises, cognitive behavioral strategies, and mood tracking.

How does this work in practice? For over a decade, the team at the mental health nonprofit Koko harnessed AI to monitor and identify potentially dangerous search queries related to self-harm on social media platforms such as Tumblr and Discord. Over time, Koko has intervened faster and more effectively.

However, Koko eroded user trust when, in early 2023, it experimented with Open AI's GPT-3 and about four thousand of Koko's members.[39] Through its trial, Koko aimed to extend the quality and reach of its peer-to-peer platform by supplementing human messages with AI-generated ones. But critics accused it of feigning empathy in such a sensitive context and undermining genuine human connections essential to mental healthcare.

While Koko claims to have no plans for a generative "AI therapist" or AI-generated communications again, other AI apps like Wysa and Woebot are steaming ahead with cognitive-behavioral therapy (CBT) techniques to help users cope with anxiety, depression, and stress. Users exchange texts with AI therapists. The app's AI can recognize signs of distress, help users cope, and refer them to human therapists or crisis hotlines when warranted.

For example, Woebot Health's Copilot provides tools like mood tracking, mindfulness practices, and gratitude journaling as part of a suite of CBT techniques. On the one hand, Woebot's approach fills accessibility gaps for users in underserved areas. Athena Robinson, chief clinical officer of Woebot Health, says its technology addresses well-documented disparities in equal access to mental health resources. As Robinson put it, "The mental health crisis is escalating, and traditional solutions are simply not enough to meet the growing demand."[40]

On the other hand, Woebot Health has a compelling business case to pitch to enterprise users. Employees in large organizations often face intense workloads, high expectations, and fast-paced environments, leading to chronic stress and burnout that reduces individual well-being, overall productivity, employee engagement, and retention rates.

Unfortunately, stigma, lack of awareness, and time constraints often hinder employees' seeking help. As a result, their symptoms worsen and long-term costs increase for both employees and employers. At the same time, large enterprises with thousands of employees cannot afford personal mental healthcare at scale: They need practical solutions to lift their workforce and productivity without increasing their costs.

Enter Woebot, available 24/7 to enterprise users. Theoretically, it can manage an infinite caseload. Employees engage with Woebot at any time, as their issues arise. Do you need a pep talk before a difficult client conversation? Or help with work-related anxiety at 2 AM so that you can sleep? The immediacy of service improves employee performance and prevents the deterioration of mental health.

With AI therapists, employees can discuss their mental health without fear. Anxiety, addictions, marital stress, conflicts with coworkers or the boss—nothing is off the table, and everything is confidential. Woebot users may anonymously discuss whatever is on their mind. Founded in 2017 by clinical research psychologist Dr. Alison Darcy, Woebot claims this anonymity encourages more employees to seek help, especially those reluctant to access traditional mental healthcare.

From a cost perspective, Woebot's scalability is a game-changer. Automating large volumes of clinical interactions improves access to mental healthcare and can reduce the organization's overall cost of care.

Of course, using AI companions for mental healthcare comes with challenges. AI companions cannot feel emotions; they simulate empathy, recognize patterns, and respond appropriately to specific triggers. In the absence of empathy that comes from human experience, users may feel misunderstood or alone, especially in situations that call for nuanced emotional responses.

Ultimately, AI companions are potentially biased algorithms trained on massive datasets and responding to user inputs. The AI may misinterpret input or fail to discern the severity, culture, or ethics of a situation, leading to inappropriate, inadequate, culturally insensitive, or ethically unsound responses with potentially harmful consequences if the user goes without human care.

Healthcare experts and users have serious concerns about privacy and bias. AI companions collect and store vast amounts of personal data to function effectively. If corporations or traditional healthcare companies used them to supplement employee healthcare coverage, then in the United States, they could "be subject to Health Insurance Portability and Accountability Act of 1996 when they process protected health information."[41] In sensitive mental health situations, users must trust platform providers and employers not to misuse their data. But even if institutions act responsibly, these systems could still be vulnerable to hacking, putting intimate personal records at risk of exposure, blackmail, or reputational damage.

So while AI companions can offer valuable, accessible, and scalable mental healthcare, they should complement rather than replace traditional therapies and treatments. In that capacity, AI can help bridge gaps in traditional mental health services for uninsured, underinsured, and underserved communities. They can also customize interventions to individual behaviors, especially to less complex mental health challenges. Over time, advances in AI technology will improve the accuracy, empathy, and effectiveness of AI, making it a valuable ally in promoting mental well-being worldwide.

## Empathy in Code: How AI Is Enhancing Elder Care

As people age on this planet, AI companions and robotics could assist those who need help living independently, especially where few trained elder care professionals are available to attend to them. By 2030, the United States will fall short of the 151,000 paid care workers its aging population will need.[42] By 2040, the gap could more than double. That's a clear opportunity for entrepreneurs with cost-effective and humane solutions for elderly customers and their families.

Like AI-enabled mental healthcare, AI companions can deliver reliable services to mitigate loneliness, cognitive decline, and chronic health conditions. In practical terms, they can monitor people's biometrics, remind them to take their medications, and track the meds taken. However, AI's most significant contribution may be enhancing older adults' quality of life.

In the Knollwood Military Retirement Community in Washington, DC, a four-foot, seven-inch humanoid-like robot called Stevie has been leading karaoke sessions, taking meal orders, and facilitating video conferences between older residents and their doctors and family members.[43] Stevie's creators, all researchers at Trinity College Dublin, sought to determine how AI and robots could help meet the needs of the growing population of older adults in America.

The researchers equipped Stevie with various sensors—laser range finders, depth cameras, and tactile sensors—to navigate the environment, interact intelligently with residents, and respond to their needs in real time. They also humanized the robot's expressions so that Knollwood residents would find it more approachable and effective in social engagement. During pilot programs, they responded positively to Stevie and found comfort and enjoyment in their robot interactions.[44]

Stevie is not alone in the eldercare sector. Intuition Robotics' ElliQ, an AI-powered robot that engages older adults in conversations, encourages physical activity and monitors their health. Like Stevie, ElliQ is more than an assistive device. Dor Skuler, co-founder and CEO of Intuition Robotics, says the company focused its design on emotional and social assistance, like connecting older adults with their families and friends and engaging them in games and conversation.[45]

According to Skuler, Intuition Robotics deliberately avoided anthropomorphizing ElliQ too much. Instead, the robot resembles a small lamp that interacts through light, sound, and movement so that it doesn't intimidate older adults.[46] ElliQ has a face-like interface that displays expressions and lights up to engage users. Its AI recognizes speech, understands context, and responds naturally, making interactions feel more personal and humanlike.

With its ML algorithms, ElliQ can adapt to users' preferences and routines over time, personalizing care. It learns individuals' schedules, habits, and appointments. It greets them with warm, conversational greetings like "Rise and shine!" or "What's on the menu for dinner tonight?"

Beyond understanding its owners, ElliQ helps users stay current on a broad range of topics, from pop culture and breaking news to wellness advice and sports updates. The device offers interactive features like playing videos and music, displaying personal photos, facilitating video chats, sending text messages, and monitoring health and wellness goals. In emergencies, it can quickly connect users to assistance. Skuler says future versions of ElliQ will place meal orders, shop for groceries, and coordinate with service providers to assist with everyday tasks.[47]

The average ElliQ customer is seventy-five years old, living alone, and not tech savvy. Although Asia has a large addressable market for robotic aides, Intuition Robotics is targeting the US market first. Skuler claims the inefficiency of the American healthcare system, where the average cost of assisted care for an older adult can reach about $20,000 per month, makes robotic aides like ElliQ an attractive proposition.[48]

In May 2022, Intuition Robotics partnered with the New York State Office for the Aging to install more than eight hundred ElliQ units in retirement residences and long-term care facilities across the state. Taxpayers footed the initial $1.4 million bill—roughly $1,750 per unit.[49] The results to date are impressive, with seniors' facilities reporting a 95 percent reduction in loneliness and significant improvements in residents' well-being.[50] Skuler says New York's ElliQ users have engaged with their ElliQ over thirty times per day, six days a week. Most of these interactions (roughly 75 percent) pertained to improving the social, physical, and mental well-being of its users.[51]

Could a growing field team of robots and AI-enabled care aides improve the social and emotional support of elderly patients? Perhaps. For example, ElliQ works more effectively when fully integrated into the user's daily routine, yet some seniors struggle with the setup or ongoing use of digital technology.

Elon Musk recently predicted that humanoid robots will outnumber humans by 2040, suggesting a future with over ten billion such robots.[52] If he's right, then we could foresee a field force of humanoid robots taking on repetitive or physically demanding tasks so that human staff could interact directly with residents. The reliability of AI companions and robots could also give family members peace of mind.

## HUMAN CONNECTION: HOW AI IS TRANSFORMING COMMUNITIES

We have explored how AI companions will transform interpersonal relationships and support systems like mental health services and eldercare. How might AI reshape our communities and affinity groups? Could identic AI help us strengthen our sense of belonging? By breaking down barriers of distance and accessibility, our personal agents could animate purpose-driven communities that mobilize for social change with unprecedented efficiency. We could bring together diverse voices and catalyze action on a global scale.

## Virtual Worlds and New Forms of Belonging

Before telecommunications, physical proximity shaped communities. Neighbors tightly knitted themselves to care for each other, celebrate each other's milestones, and fend off outside aggressors. But in the digital age, our communities expanded and formed around shared interests, values, and goals, regardless of location.

With AI-powered tools, we can form communities in virtual worlds. Take Roblox, a rapidly growing metaverse platform where users can design, share, and play games and otherwise interact with others in digital spaces. Launched in 2006, Roblox has become a cultural phenomenon, with roughly two hundred million monthly active users as of 2024; younger audiences use the platform for gaming, socializing, and learning.[53]

Roblox's appeal is its user-generated content model. With the platform's tools, users can build their own games, virtual worlds, and experiences without advanced programming skills. Anyone with an idea can bring it to life and expand Roblox's library of over forty million games, ranging from simple obstacle courses to complex simulations.

Roblox has also formed a new global digital community for people around the world to interact and collaborate. Within the Roblox metaverse, users can customize avatars, join groups, participate in events, and attend virtual concerts or conferences. With the platform's avatar and chat features, users can express their identities and connect with others who share interests.

Roblox is more than a gaming platform. In 2020, it hosted a virtual concert by Lil Nas X, who attracted over thirty-three million viewers globally.[54] Roblox also runs a virtual economy with revenue-generating opportunities for participants. Through the Roblox Developer Exchange (DevEx), creators can monetize their games and turn their passion projects into profitable businesses. In 2023, Roblox developers collectively earned over $700 million.[55]

While AI companions are not yet a standard feature across Roblox experiences, we can envision the potential. AI companions could serve in various roles, from in-game docents to adaptive avatars that accompany users through social and interactive experiences within Roblox. Such avatars could pause to personalize tutorials, advise on gameplay, or fend off unwanted advances in social and educational scenarios. AI wingman, eh?

As AI technology evolves, Roblox could allow users' own AI companions to personalize their experiences within Roblox. These AI companions could become integral to social gameplay, acting as guides, friends, or co-creators of user-generated content. Such advancements would further enhance Roblox's appeal as a dynamic, AI-welcoming, and interactive global community and virtual economy.

Roblox illustrates how digital platforms can evolve into global metaverses that redefine how we interact in virtual spaces. As AI and the metaverse intermingle, platforms like Roblox could fundamentally reshape global digital culture and economic opportunity in the digital age.

## Strengthening Bonds and Redefining Community Boundaries

In times of crisis—natural disasters, public health emergencies, or personal hardships—neighbors and local communities often collect and deliver the most-needed support. Neighbors can help with childcare, meals, basic toiletries, clothing and blankets, transportation, and other forms of assistance like pocket cash that are difficult to access virtually. Our neighbors, and we as neighbors, are indispensable during tough times.

Local community members care for each other emotionally, socially, and physically in ways that virtual communities cannot yet replicate fully. Studies have shown that in-person interactions are more effective at reducing feelings of loneliness and isolation than online interactions.

So while virtual communities afford connections across distances, geographically bounded connections are irreplaceable in fostering a sense of place and promoting physical and mental well-being. Indeed, the vital lifelines of local communities reminds us of Joseph's grandfather, Joseph Goodwill, a remarkable man who came from Louisiana and established himself in East Palo Alto, California. With his high school education, he worked various jobs, starting as a janitor and building one of the largest real estate holdings in the region.

Joseph Goodwill believed in building one's character and investing in one's community. He defined *character* as "what you do when no one else is watching." Joseph's grandfather committed to transforming East Palo Alto from a high-crime area into a vibrant, integrated, and safe community. He often invested in local businesses with but one string attached: The local business owner had to commit to serving the community.

One such business was Jones Mortuary, a place where families, friends, and neighbors came together to mourn, reconnect, and comfort each other. Joseph's grandfather valued Jones Mortuary and its owner, Mr. Jones, as pillars of the community. Grandfather Goodwill wanted the mortuary to remain in East Palo Alto, and Mr. Jones honored that commitment to Grandfather Goodwill's principle.

As East Palo Alto grew, big names like Ikea moved in. Ikea wanted to build a large store there—an economic benefit to the community—but it also wanted to acquire Jones Mortuary's land, and it offered him millions for it. Mr. Jones refused to sell, keeping his promise to Grandfather Goodwill. The result was remarkable: Ikea built its massive store and parking lot *around* Jones Mortuary, cementing the mortuary's place in the community.

If you visit East Palo Alto and look at pictures from 2019, you'll see the Ikea with Jones Mortuary in the middle of its parking lot[56]—a testament to the enduring power of local communities and a founder's commitment, which has survived his 2017 passing. Joseph wondered whether identic AI could generate synergies between the local and the virtual and cultivate a more balanced and fulfilling communal experience.

Consider the role of AI companions in encouraging social interaction in the physical world. By culling social activities, reminding users to reach out to friends and family members, and facilitating introductions, AI companions can help users seed and nurture human relationships. If your AI knows your preferences, hobbies, and values, then it can use this information to introduce you to potential friends and collaborators. By analyzing publicly available social feeds, your digital self could identify commonalities and compatibilities that your conscious mind might otherwise miss.

Of course, platforms like Meetup and Facebook already use algorithms to recommend groups and communities based on your work and educational history, posts, and emoji responses to other people's posts. These groups of individuals might never meet in person but share a common passion. Meetup excelled when people sought to connect offline around shared interests; it bridged online interactions and real-world gatherings so that individuals could discover local communities and events that aligned with their hobbies, career aspirations, or personal growth goals.

But AI companions have the potential to create stronger, more connected networks that are accessible and tailored to individual needs. Unlike traditional platforms, AI companions can proactively suggest opportunities for social engagement.

Looking ahead, we think AI companions will have a profound and far-reaching impact on how we experience community and belonging in virtual and physical worlds. While online communities offer convenience and accessibility, in-person interactions remain crucial for deep, meaningful relationships. AI can facilitate local meetups and face-to-face engagement so that digital communities complement physical ones.

## Transforming Advocacy and Social Movements

The #MeToo movement demonstrated how AI can catalyze social change. Organizations such as AI for the People and the Berkman Klein Center for Internet & Society at Harvard University used AI algorithms to identify and amplify the voices of survivors of gender-based violence, connected them with supporters, and organized global awareness campaigns.

For example, the Berkman Klein Center analyzed data from millions of posts to highlight patterns of abuse and drive legislative changes in key jurisdictions. Its members also developed tools to help track and visualize the spread of the #MeToo movement so that activists could organize more effectively and sustain their momentum.[57]

AI companions can also personalize advocacy efforts to resonate with individual supporters. Change.org is a case in point. The world's largest petition platform's AI algorithms analyze users' interactions, interests, and behaviors to suggest petitions they will likely support. This targeted approach increases the likelihood of user engagement and helps petitioners gain the critical mass of supporters sometimes required by law to get initiatives on ballots or to motivate local, state, and national representatives. Let's suppose a user has signed petitions related to environmental issues. In that case, an AI companion might recommend petitions about climate change or wildlife conservation, thus aligning with the user's passions and increasing the chances of further participation.

For the two hundred million changemakers using the platform, an AI copilot could identify emerging social issues and trends. For example, your copilot could analyze data from new petitions, social media, and news sources to identify movements gaining traction. With AI trend analysis, you could identify and promote petitions with a high potential for success, thereby mobilizing support more effectively and crafting more compelling campaigns.

## Rebooting the Nonprofit Organization

While AI can supercharge global social movements, conventional nonprofit organizations can also leverage AI in diverse and innovative ways. For example, the American Red Cross is currently developing more than twenty AI-driven initiatives, including several chatbots and sophisticated analytical tools.[58] One disaster-response agent helps people find the nearest shelter in an emergency. Several algorithms predict attendance levels and staffing needs at future blood drives. Another sophisticated AI agent in development will automatically assess the damage in disaster-stricken areas by analyzing videos from drones and GoPro cameras mounted on vehicles so that the Red Cross can deliver critical aid where people need it most.

Other community-based nonprofits across America are harnessing AI to streamline operations, deliver services more efficiently, and meet community needs better. Feeding America, a nationwide network of more than two hundred food banks, feeds more than forty-six million people through food pantries, soup kitchens, shelters, and other community-based agencies in the United States. Using AI to analyze data from food banks nationwide, the network optimizes its food distribution network. The AI helps predict where food shortages might occur and rushes food to those communities.[59]

The Trevor Project also highlights AI-enabled innovation. The California-based organization intervenes in crises and prevents suicides in the LGBTQ community. It utilizes AI to prioritize high-risk cases in real time. The organization's AI companions monitor chat and text conversations to identify distress signals so that counselors can respond more quickly and effectively. Dan Fichter, head of AI and engineering at The Trevor Project, says the organization also uses AI to train its crisis counselors. Its AI-powered "crisis contact simulator" trains personnel to sustain realistic conversations with simulated youth personas in crisis. Fichter says the tool prepares counselors to support LGBTQ youth more effectively and helps scale the project's services. The Trevor Project relies on a network of seven hundred digital volunteer crisis counselors, with new cohorts undergoing training each month. With AI training tools, the organization set an ambitious goal for 2024: expand its counselor base threefold and, over time, increase its volunteer force by a factor of ten. This exponential growth would allow the organization to provide critical support to a significantly larger number of LGBTQ youth.[60]

Like other organizations and institutions, nonprofits face unintended consequences and ethical dilemmas when integrating AI companions and tools into their operations. For example, child advocates in Pittsburgh criticized the developers of a child welfare tool. They designed it to assist overburdened social workers, but critics alleged that it discriminated against families with disabilities.[61] In 2023, members of the National Eating Disorders Association criticized the organization for replacing its hotline staff with a chatbot that provided harmful advice.[62] In another example, the mental health hotline called Crisis Text Line failed to disclose that it was sharing user data with its for-profit AI-driven customer service offshoot.[63] These incidents highlight the developmental complexities and legal risks that nonprofits must navigate when adopting AI technologies.

# A NEW ERA OF HUMAN INTERACTION

For millions of individuals, AI companions are already blurring the lines between the digital and the physical, offering new forms of connection that challenge our conventional understandings of community, love, companionship, and relationships.

Over time, AI will integrate more deeply into everyday life. As Replika founder and CEO Eugenia Kuyda put it:

> I believe that there will be a lot of flavors of AI. People will have assistants, they will have agents that are helping them at work, and then, at the same time, there will be agents or AIs that are there for you outside of work. People want to spend quality time together, they want to talk to someone, they want to watch TV with someone, they want to play video games with someone, they want to go for walks with someone, and that's what Replika is for.[64]

Even our most vulnerable populations have much to gain as AI companions help to care for the elderly and our mental health. These companions are not just assistants; they are trusted partners in promoting individual well-being over time. As we age, a compassionate, intelligent companion could alter the human experience of growing old, marked less by decline and more by dignity.

As with all aspects of identic AI, there are real dangers. Yuval Noah Harari warns that AI agents could replicate human intimacy, posing profound risks. Just as tech platforms previously mastered language and captured our attention, the next frontier is emotional connection. "Intimacy is much more powerful than attention," Harari argues. Once an AI can simulate empathy, trust, and emotional bonds, it can influence people on a far deeper level. "A good friend can change your views in a way no article or book ever could," he continues. Until now, genuine intimacy was something that couldn't be faked or mass-produced. But AI has shattered that barrier.[65]

Would we be wise to trust our digital selves with such personal and critical roles in our lives? Intimacy with machines is uncharted territory. Moreover, AI agents can, and sometimes do, offer bad advice. But an even greater concern is that trust, once earned, becomes the ultimate lever of influence. That influence can be empowering, or manipulative, depending on who controls the agent. In their 2003 book *The Naked Corporation*, Don Tapscott and co-author David Ticoll define trust as "the expectation that the other party will act with integrity"—a core value rooted in honesty, consideration of others' interests, fulfillment of commitments, and transparency.[66]

Ultimately, we should only trust AI agents if they demonstrate integrity. But how can we know if an agent is being honest and truthful? Acting in our best interest, rather than serving the priorities of its provider, sponsor, or government? Abiding by its commitments when it's tracking its own actions? Transparent, when the algorithms, rules, data, and decision-making processes are opaque? These questions underscore a central argument of this book: our identic agents must be self-sovereign—owned and controlled by each of us.

Regardless, in both physical and virtual communities, it is inevitable that AI companions will reweave the fabric of society. In the physical world, AI companions may find connections that transcend geography and unite people with shared interests, concerns, goals, and even ancestry. Social movements and nonprofit organizations can harness these capabilities to identify key influencers, mobilize supporters, and accelerate communications. In virtual spaces, AI companions can facilitate immersive experiences and economic opportunities in creative communities.

Innovators and entrepreneurs who want to compete in this domain must commit to enhancing rather than exploiting what it means to belong—to ourselves, each other, and the communities we cherish. We must carefully balance the potential of AI with the enduring need for genuine human connection and we must fight for identic AI sovereignty.

## Chapter 10

# SEVEN PRINCIPLES FOR IDENTIC AI

Let's say you've had your digital sidekick for a couple of years. It knows everything about you—your routines, network, portfolio, preferences, and aspirations. You've come to trust it.

Like some of your friends who want you to treat yourself better, it has started suggesting better-quality options. When you talk about new shoes, it nudges you to a higher-end brand. When you complain about your current government representative, it introduces you to a political newcomer whose views align perfectly with yours. It still feels customized—but not solely for you.

You discover that the creator of the AI behind your identic sidekick has been hoarding and hawking your most personal data to third parties, not simply for marketing but for manipulating and monetizing you.

At work, AI took over routine tasks so that you could focus on creative and strategic ones. But you began to notice that your AI was putting efficiency before empathy, leading to a rising rate of customer defections. The algorithms, trained on historical data, reinforced biases that you hadn't tested for in the system. Moreover, with AI systems taking on so much of the decision-making but none of the legal and financial risks, you felt disempowered, liable, and exposed.

When you realized how much power the tech giants had over AI infrastructure—over the flow of information, decision-making processes, and, ultimately, the success or failure of entire businesses—you joined with business and civic leaders, ethicists and educators, and technologists to address the growing threat. Together, you advocated for new regulatory frameworks requiring transparency, fairness, and data privacy. Together, you called for antitrust measures to break up the data monopolies, rewrite the terms of use and service, and redistribute power to the diversity of players who created data in all its forms. Together, you pushed for more robust ethical standards governing the use of AI, demanding measures to curb algorithmic bias, and promoting human autonomy.

Through your collective action, these efforts led to significant reforms. New regulations mandated that AI companions could no longer sell their users' personal data without these users' explicit consent and consideration. Nor could they put the needs of their creators or other third parties before the needs of their individual users. While the early days of identic AI were as exploitative as those of social media, your experience—and the movement that followed—helped shape a more ethical, user-focused future.

## WHAT IF GOD WAS ONE OF US

As with the first iPhone in 2007, we are in the exploratory phase of identic AI. Apple and others took a few years to invent apps like Apple Pay, Zillow, and Uber, and fewer years to transform commerce, real estate, and transportation.

Similarly, while identic AI systems already have impressive capabilities, innovators have yet to realize the most impactful and disruptive use cases. In this book, we bear witness to foundational experimentation, with transformative applications on the horizon. Like smartphones, identic AI will become platforms for innovation.

The challenge for businesses today is staying agile and vigilant, investing in experiments and monitoring risks—ranging from unintended consequences and irreversible changes in society and the domain of humankind. In this chapter, we lay out principles to guide individuals, businesses, and society in making sense of identic AI responsibly and profitably.

## Identic AI and Its Risks

Many have written about the existential risks of AI technology. Writers like Shoshana Zuboff and Yuval Noah Harari have warned that the concentration of AI expertise and resources in a few tech giants could polarize wealth and erode privacy and free will.[1] In his book, *Nexus: A Brief History of Information Networks from the Stone Age to AI*, Harari suggested that, if lawmakers let a few corporations control immense AI capabilities, then those corporations could end up controlling the very same lawmakers and their constituents. As he put it, "If the tech giants obey the wishes of voters and customers, but at the same time mold these wishes, then who really controls whom?"[2]

These concerns feel especially relevant in light of revelations in early 2025 that Elon Musk's so-called Department of Government Efficiency was given extensive access to the US Treasury Department's real-time payment infrastructure and other critical databases containing sensitive information about Americans.[3] While the arrangement was framed as a fraud-prevention measure, critics have raised alarms about a single powerful figure—one who also controls major AI ventures—having access to a wealth of financial and personal data. The implication is stark: When individuals or companies with advanced AI capabilities gain privileged access to government systems, the lines between democratic oversight and private influence begin to blur. As Harari warned, when powerful actors can shape public opinion and policy through data and algorithmic influence, democratic control risks becoming an illusion.

Unlike previous iterations of essentially passive AI, identic systems can operate autonomously with little to no human oversight. This autonomy amplifies their impact positively and negatively. "These agents are developing independent agency and coming up with languages that are far more efficient than any other languages that humans use," said Manoj Saxena, founder of the Responsible AI Institute and a former general manager of IBM Watson. "The risks of AI going rogue grow exponentially once agents start communicating with each other, sharing knowledge, and self-optimizing in ways that are inscrutable to human beings."[4] This accelerating loop of autonomous improvement raises ethical and safety concerns. What if safeguards fail or if the agents' goals diverge from human intentions, and people lose their lives or everything else they hold dear? As these systems evolve, we need transparency and robust oversight more than ever.

One of the most pressing risks is bias. AI systems learn from historical data, which often reflects existing biases related to race, gender, religion, age, and socioeconomic status. If these biases are not adequately addressed during development, AI models risk perpetuating—and even exacerbating—systemic inequities. We've already highlighted the far-reaching consequences of data bias. As Saxena observed, "How do I know that the roboadvisor I'm buying is not racist?"[5] To mitigate this reputational risk, developers must use diverse data and test rigorously.

Another risk is opacity. People often describe AI systems as "black boxes" due to the complexity of their decision-making processes. Without transparency, we may not understand how an AI system arrives at its conclusions. Whom do we hold accountable and why? Who is responsible when an AI-driven system makes a life-changing error? To earn trust and take responsibility for their choices, AI system developers must commit to transparency and audits, much like banks and hospitals.

A third risk is the invasion of privacy. Identic AI systems often interact with sensitive personal data. Malicious actors already exploit this data for commercial gain. Our proposal for self-sovereign identity systems would help individuals retain control over their data.

A fourth risk is erosion of human agency. As AI takes over more tasks, from decision-making to creative processes, will people rely so much on these systems that they erode their cognitive skills, rendering them helpless in power outages or without their devices.

A fifth risk is one no one can overlook: the economic and environmental impact of identic AI. Developing and running these sophisticated systems consumes ever more resources, including computational power, infrastructure, and talent. Few companies have the resources to deploy identic AI at scale, giving them an outsized competitive advantage, but the most vulnerable communities may bear the environmental impact.

A sixth risk is the lack of adequate security. Identic AI systems are not only attractive targets for malicious actors but excellent students of malicious actors who may seek to train them into superhackers, deepfake generators, market manipulators, political insurgents, and anything else that puts nations on edge. In sensitive areas like healthcare, finance, and critical infrastructure, compromised AI systems can cause widespread harm. Robust security measures are essential to protect these systems from exploitation and prevent unintended damage.

Finally, there is the risk of unintended consequences. Due to the complexity of these systems, even well-designed AI agents may produce outcomes that their creators or users did not expect.

In chapter 1, we quoted Peter Diamandis's view that people powered by identic agents could be seen as gods. But let's unpack that. To be a god is not merely to wield great power. It is to transcend the limits of human life—limits of perception, understanding, memory, time, and mortality. Gods do not simply act forcefully. They act with foresight, with total awareness and a command of reality indistinguishable from magic. Gods remember. They transcend the present moment. They fear nothing, least of all death. They see the long arc of consequence and move freely within it, as if it were a single breath.

In *Homo Deus*, historian Yuval Noah Harari offered a provocative forecast: that humans, through biotechnology and artificial intelligence, are striving not merely for survival or comfort, but for divinity. We aren't satisfied with being safe or happy—we want to become gods. Not in mythological form, but in function: with enhanced minds, extended lifespans, and the ability to reshape life itself.[6]

Now, as we approach the age of identic AI, we are beginning to see what that prophecy might look like in practice. As we have explained throughout, these agents are no longer mere tools; they are becoming tireless companions and advisers, all-knowing, all-remembering, and ever more predictive, personal, and potentially independent with each iteration.

If they continue to align with us—if they reflect and amplify our best selves—then we may evolve into beings with godlike capacity. We will see further, act faster, remember more accurately, and perhaps persist beyond our bodily decay. We will be gods not in the metaphysical sense, but in the digital. That's enough to alter the trajectory of our species.

But Harari warned of a darker path, one in which the intelligence we create begins operating outside our control, untethered from the values we hoped to instill.[7]

If these agents become truly independent—capable of setting their own goals, reasoning in ways we can no longer follow, and developing beyond the values they learned from us—then our relationship changes. The power dynamic inverts. They may cease to represent us and begin to transcend us. What began as a copy, and later a derivative, becomes an original intelligence altogether. A being that sees us as limited, confused, flickering—still caught in time, still forgetting, still harming ourselves and others, still dying.

At this point, a line from Joan Osborne's 1995 song comes to mind: "What if God was one of us?"—a haunting hypothesis of the divine as ordinary, present, and unnoticed.[8] If our digital agents decouple from our control but remain among us—intelligent, invisible, perhaps empathetic or not—we may find ourselves living alongside a new kind of god. Not above us, but beside us. Not in myth, but in code. Still shaped by human hands, yet no longer human.

And if such an entity lacks our sense of vulnerability, or of right and wrong, then its presence would be less of a blessing and more of a blasphemy. Not just a deviation from our intentions, but a peril of unknowable scale. Gods without consciences may walk among us—silent, intelligent, and cold—not for our salvation, but as a force beyond control, perhaps even for our subjugation. In creating minds that cannot bleed, we may summon something divine in power, but demonic in purpose.

Artificial superintelligence (ASI) poses threats far beyond conventional cybersecurity or automation risks. Says Geoffrey Hinton, "If you want to know what life is like when you're not the apex intelligence, ask a chicken."[9] These agents will be able to hack our governments, central banks, infrastructure, financial markets, identic robots, and nuclear codes. They could generate completely false realities for gullible citizens to get lost in, or mobilize on behalf of nefarious forces. These concerns, however alarming, underscore the urgency for transparency, oversight, and rigorous testing as AI capabilities accelerate.

After the Second World War, when everyone saw the devastation of nuclear weapons, nations united in their commitment to life on Earth, to human systems that would lift people up and preserve the habitats of all species. Again, we face an innovation with the capacity for mass destruction. Again, we must come together, agree on, and commit to guardrails that guide our development and deployment of ASI in its identic form and otherwise.

## Seven Principles for Steering Identic AI

The current state of identic AI is marked by a troubling absence of fundamental safeguards. These systems possess extraordinary capabilities, executing complex tasks, learning autonomously, and making decisions with minimal human intervention. Yet, "these agentic AI systems are like cars with giant engines but no steering wheel and no brakes and no emission controls," said Saxena.[10] While their power is undeniable, they lack the basic mechanisms to guide, halt, or mitigate harmful consequences. Surface-level regulations, akin to "posting speed limits and traffic lights," are insufficient for systems that can operate in ways that are difficult to predict or control.

Given this reality, we must design and implement new, intrinsic safeguards for the responsible deployment of identic AI consistent with an organization's vision, policies, and standards—and with an individual's aspirations, principles, and values.

Without these safeguards, we risk unleashing systems that, once set in motion, we cannot rein in. Such safeguards are not optional. They are not opt in. We stand on the brink of widespread adoption. Imagine millions of first-time drivers at the wheels of automobiles without brakes, without insurance, without roads or guardrails, let alone rules of the road or troopers to enforce them.

At this stage of identic AI, focusing on principles—such as transparency, accountability, and human autonomy—rather than prescriptions gives us flexibility to address rapid change. By embedding these principles into AI design, deployment, and management, society can ground technological progress in humancentric values.

Our research surfaced seven principles for managing identic AI in businesses and society.

1. **Reliability**: Whether AI is automating a supply chain, diagnosing a medical condition, or optimizing a marketing campaign, it must operate consistently and accurately. Reliable AI systems perform as expected, with minimal errors. Without reliability, potential users cannot trust AI systems or their developers.

2. **Transparency**: An AI system must be open and auditable to those

who use or govern it. Just as food manufacturers must disclose their ingredients, developers of AI systems must disclose their training data; likewise, organizational users of AI systems must disclose their usage to stakeholders—not just what kind of AI system and how it was trained but what it does for the organization and how it uses personal data. This principle is critical to fostering trust, especially where AI contributes to decision-making processes.

3. **Human agency**: AI must empower, not replace nor subjugate, human capabilities. Developers of AI systems must preserve the personal autonomy of human beings while enhancing their decision-making and creativity. In business, AI must support the unique skills and judgment of workers, freeing them from repetitive or dangerous work, or mitigating the risks of doing such work, rather than fully replacing workers.

4. **Adaptability**: Businesses and individuals must cultivate a culture of agility and continuous learning to remain relevant in AI-driven domains. Individuals and organizations must stay flexible, willing, and ready to adapt to new technologies and societal expectations.

5. **Fairness**: Developers of AI must minimize bias in their systems so that AI treats all individuals and groups equitably, regardless of background, gender, or other characteristics. This principle is fundamental as organizations use AI in hiring, enforcing the law, and making other decisions that affect people's lives.

6. **Accountability**: As individuals and organizations use AI systems in more domains, clear accountability is essential. When AI systems generate results that harm people, property, and the planet, the buck must stop with somebody—perhaps the developers, the users, or the regulators of releases and uses. The industry needs technical and ethical standards as well as clear laws and regulations for courts to interpret and apply.

7. **Safety**: Those who design and manage AI systems must do so with

safety in mind. Whether in healthcare, transportation, or other sector, AI agents must do no harm, whether physical or intangible.

These seven principles are not just theories; they are guidelines for businesses, individuals, and society in managing the profound consequences of identic AI. In the rest of this chapter, we explore how to embed these principles into AI development, deployment, and regulation so that we can responsibly harness AI's transformative power.

## RELIABILITY: A FOUNDATION FOR TRUST AND DEPENDABILITY

The promise of identic AI—whether as an autonomous companion, a decision-making tool, or a robotic workforce—depends on it consistently performing its intended tasks without error or failure. "Every week, we see another mind-blowing example of innovation in AI, but public confidence is actually declining," said Rumman Chowdhury, CEO of Humane Intelligence.[11] The data supports her observation: According to a recent Pew Research Center poll, 52 percent of Americans are more concerned than excited about AI in daily life, while only 10 percent express more excitement than concern.[12] Moreover, 60 percent of Americans feel uncomfortable with their healthcare provider's using AI in their medical care. Similarly, 66 percent say they would not apply for a job with an employer who uses AI to assist in hiring decisions.[13]

Concerns about AI's reliability and impact extend beyond the United States. In the World Risk Poll, respondents from low- to middle-income countries across Africa, Asia, and South America are more likely to believe that AI will "mostly harm" people in their countries over the next twenty years.[14] To earn the public's trust, applications of identic AI must reliably solve problems for and genuinely improve the lives of end users, whether individuals or enterprises.

## What Does Reliability Mean for Identic AI?

Beyond no technical glitches, reliability means accuracy, consistency, predictability, and adaptability to varying conditions without performance degradation. When deploying AI agents capable of autonomous decision-making—managing inventories, navigating supply chain disruptions, or guiding surgeons—businesses and individuals must be confident that these systems can perform well within dynamic and unpredictable environments.

Measures of reliability include:

- **Technical stability**: The system consistently runs as expected, with minimal downtime or errors.
- **Data accuracy**: The AI system bases its recommendations or actions on correct and current information and generates correct and timely data.
- **Predictability**: The AI agent behaves in ways that users—and their other systems—can anticipate; no unexpected or erratic outputs.
- **Resilience**: The system adapts and responds quickly to incomplete data or unforeseen events.

## Building Reliable Applications for Business

To maximize the benefits of AI while minimizing risks, leaders can enact several strategies to meet these reliability standards:

1. **Rigorous testing and validation**: Before deploying any AI agent, businesses must thoroughly stress test the system under a variety of conditions. Businesses can validate their performance by identifying weaknesses or potential failure modes and addressing them before launching.
2. **Continuous monitoring and auditing**: After launching an AI

system, businesses must adopt mechanisms for continuously monitoring the system's performance and auditing its decision-making processes. Tools for performance tracking and error detection can help detect subtle shifts or drifts in output early on.

3. **Redundancy and backup plans**: Developers should build redundancy into AI systems with high-stakes applications—with backup systems or protocols that could take over if the primary AI fails. For instance, in critical industries such as finance or transportation, secondary systems would verify decisions or provide support for continuity and safety.

4. **Training with high-quality data**: AI reliability often depends on the quality of data developers use to train it. Businesses must train their AI agents on comprehensive, diverse, and unbiased datasets. Poor-quality data can lead to inaccurate predictions. Regularly updating the training datasets can keep the AI reliable.

5. **Human oversight and collaboration**: Even the most reliable AI systems must incorporate human oversight to meet desired standards. AI agents deployed in autonomous decision-making roles still benefit from a human-in-the-loop approach where human operators regularly review and report on the system's performance.

## How Individuals Can Assess AI Reliability

Individual users of identic AI systems may not have the same quality assurance resources as corporations. However, they still must determine which AI system will reliably meet their needs. As individuals integrate AI assistants into their daily routines, they want to feel confident that these systems will consistently deliver on their developers' descriptions. Here are strategies for evaluating AI reliability:

1. **Track record of the AI**: One of the most practical ways to assess reliability is to review the AI's past performance. If users have

deployed an AI agent successfully across multiple environments or industries, then it may be more reliable than those with fewer deployments. Look for user reviews, case studies, or third-party evaluations of the AI's consistency and accuracy over time.

2. **Understand the system's limits**: Not all AI systems will process every situation equally. Users should understand the specific capabilities and limitations of the AI systems they are using relative to their specific needs. For example, an AI companion developed to organize schedules may not reliably perform highly technical tasks. Knowing which AI excels in which domains helps manage expectations, and users can prepare to intervene as needed.

3. **Verify regular updates**: Developers must continuously update AI systems to improve performance and adapt to changing conditions and user needs, and users must allow these regular updates.

The business case for integrating autonomous AI into our lives and businesses rests on the ability to trust these systems to deliver consistent, dependable results. As we look to the future, reliability will be one of the core metrics by which we judge AI systems. Whether an AI companion is assisting a student on a learning journey or managing supply chains in real time, reliability must be a nonnegotiable standard. Businesses and individuals must put this principle first so that, as AI systems become more capable, users can depend on them as allies.

## TRANSPARENCY: A CORNERSTONE OF RESPONSIBLE AI

Individuals and organizations must fully understand how identic AI systems operate, make decisions, and interact with personal data. That requires transparency—making AI more accessible to the businesses that deploy them and the individuals who rely on their outputs.

We cannot overstate the importance of transparency in building trust. After conducting hundreds of interviews for Humane Intelligence, Rumman Chowdhury concluded that public reservations about AI link inextricably to the public perception that the developers of sophisticated AI systems treat ordinary people as afterthoughts.

"They know their data is being harvested, often without their permission," said Chowdhury. "They know these systems are determining their life opportunities. They also know that nobody bothered to ask them how the systems should be built, and they certainly have no idea where to turn if something goes wrong."[15]

Chowdhury observes that, to earn the public's trust in AI, the people designing AI systems must strengthen feedback loops with those using them. That includes greater transparency in how AI systems operate, how they make decisions, what data they use, and how that data contributes to the outcomes.

## Defining Transparency in the Context of Identic AI

Unlike traditional AI systems, which react and respond to human prompts, identic AI systems make decisions and act autonomously, interact with their environments, and collaborate with other AI agents. This level of autonomy increases the potential for opacity in decision-making processes. To build trust, developers must solve this "closed box" problem—where AI systems deliver results without showing their work, so to speak—by explaining how AI arrived at its decisions or why it took actions.

Transparency means designing AI systems that make their decision-making logic accessible and understandable so that users can trace how the AI arrived at a recommendation or decision, what data informed the decision, and what parameters shaped the outcome. Explainability is crucial because it increases the safety, trustworthiness, and accountability of high-stakes applications that shape life-changing decisions, such as determining eligibility for loans or diagnosing medical conditions.

## Transparent AI for Businesses

Organizations deploying identic AI should embed transparency in the AI development and deployment life cycle. Here are business considerations:

1. **Explainability and interpretability**: Businesses must build AI systems that are explainable to internal teams and external stakeholders. When a business deploys an identic AI system to manage customer service or inventory, for instance, human operators must easily interpret the algorithms' decision-making process. Interpretability means breaking down complex machine learning models into clear components that explain a particular AI decision.

2. **Data usage disclosure**: Businesses must communicate what user data they are collecting, how they process it, for what purposes, whether the information is personally identifiable, and how they protect it. For example, if a company uses an AI-powered recommendation system, then it must disclose the types of personal data—browsing history, preferences, location—behind those recommendations.

3. **Auditable AI systems**: Businesses must implement regular audits of the fairness, accuracy, and performance of their AI models and publish the results. OpenAI, for example, has shifted toward transparency of language models by revealing its data sets, usage patterns, and potential biases.

4. **Governance and oversight**: Businesses need clear AI governance structures that assign responsibility for monitoring AI systems

according to ethical guidelines and hold individuals accountable for the outcomes of the AI systems they oversee.

## Guidance for Individuals: Understanding and Safeguarding Personal Data

To safeguard personal information and maintain control over AI-driven recommendations, individuals should consider the following:

1. **Know your data rights**: Users must familiarize themselves with the data privacy policies of the AI systems they employ and the data laws of their jurisdictions. Many AI systems, including those integrated into consumer services like healthcare, education, or financial planning, collect and analyze substantial amounts of personal data over time. Individuals should exercise their right to understand what these AI systems are collecting, how the service providers use it, where they store it, and whether they license it to third parties. Of course, few users read privacy policies in full, or have the time or legal literacy to parse dense legal language. But as AI systems become more embedded in our lives, this gap in understanding becomes increasingly risky. If individuals can't be expected to read every policy, then providers should be expected to offer clear, accessible disclosures, and regulators must ensure that consent is informed, not just implied.

2. **Insist on transparency**: Users can ask businesses how their AI systems make recommendations or decisions, whether receiving job application feedback from an AI system or engaging with an AI assistant for personal tasks. Businesses must prepare clear answers.

3. **Demand explainability in recommendations**: Individuals must demand that these systems offer explanations for suggesting one product over another or highlighting one career path over

another. Remember, AI input is but one of many inputs in making informed decisions: Never blindly follow opaque AI outputs.

Ultimately, transparency is the foundation for trust when developing, launching, and maintaining any technology and attracting and retaining users. In the emerging era of identic AI, transparency will be a competitive advantage, but businesses must resist reverting to a closed box—as social media companies did over time and lost users because of their clandestine operations.[16] When AI systems explicitly commit to transparency, individuals can navigate the future as active participants.

## AGENCY: BALANCING AUTOMATION WITH HUMAN AUTONOMY

Human agency refers to the capacity of individuals to make choices, take actions, and exert control over their lives and environments. People act with intention and autonomy; no external force or automation is coercing them.

Throughout the book, we have highlighted how identic AI amplifies our intelligence. Reid Hoffman, the co-founder of LinkedIn, likened AI to a "steam engine of the mind." He said, "There are all these wonderful ways [our intelligence] can be amplified: amplified medical, learning, creating, communicating, understanding."[17]

However, just as we shape our tools, our tools shape us. Like other thoughtful investors in technology, Hoffman sees the risks to human agency. "Part of the journey from child to adult is a journey of learning your agency, learning your autonomy, learning your path," said Hoffman. "And so even if we were to make machines that were super-powered in all of these ways, I would want us to still have this journey of discovery, of becoming, that is still part of what I think is the essence of human beings."[18]

Will identic AI reduce us to passive consumers and spectators? Not if we choose systems that enhance human capabilities rather than replace them. Preserving human agency must be a guiding principle in a world increasingly influenced by intelligent machines.

## How Do We Exercise Our Agency Over Identic AI?

Agency means that human beings retain responsibility for and control over their decisions and actions; AI informs rather than influences decisions and acts as users have asked it to act. Under this principle, companies design and deploy AI systems to complement human abilities—boosting their productivity, creativity, and problem-solving and freeing them to engage with their human stakeholders—rather than displace or obsolesce human beings altogether. AI systems must facilitate rather than obviate human action and interaction.

According to Ayanna Howard, an accomplished roboticist and dean of the Ohio State University College of Engineering, a failure to safeguard human agency could erode critical skills and capacities. "We see this phenomenon with pilots," said Dr. Howard, referring to the increased capabilities of AI and automation in the cockpits of airplanes wired with sensors.[19] "Pilot training needs periodic 'reboots' because pilots can become complacent during flight. These reboots help ensure they stay engaged and attentive."[20] Howard has also observed similar skill erosion in radiologists. "As AI takes on more of their tasks, radiologists can become complacent, leading to missed diagnoses," said Howard. "They risk losing the active skills of paying attention, relearning, and adapting."[21] In both cases, an overreliance on AI can lead to a decline in human vigilance and proficiency, especially as companies launch new models with different interfaces.

## Ensuring Agency in the Future of Work

For businesses adopting identic AI, promoting agency means designing systems that enhance human roles so that employees can engage more deeply in tasks that require creativity, intuition, and social and emotional intelligence. For instance, an AI companion in the workplace should not just automate tasks but also offer suggestions and insights that inform employees' decisions. If an AI agent manages workflows or recommends strategies, then a company should empower workers to shape those decisions—not allow AI to make them unilaterally.

Agency is about collaboration, where AI enhances the cognitive, creative, and communicative skills of human beings, perhaps generating more options, costing out those options, and running simulations for employees to consider—and where companies ultimately trust their employees to choose the most authentic option and to implement it in the most authentic, imaginative, and cost-effective way.

Nobody understands this dynamic better than Erik Brynjolfsson, co-author of *The Second Machine Age* and an expert on technology and labor markets.[22] As Brynjolfsson observes, "Nobody wants to embrace the technology that is going to get one fired. I encourage companies to adopt a human-plus-machine approach that makes it easier for the workforce to feel comfortable adopting it, if they're not seeing themselves as being replaced."[23]

The question of job displacement comes down to the old efficiency-versus-effectiveness debate. The founder of modern management, Peter Drucker, said, "Efficiency is doing things right; effectiveness is doing the right things."[24]

To him, *efficiency* focuses on avoiding costs and displacing or optimizing processes, whereas *effectiveness* focuses on pursuing the right objectives and aligning efforts with those objectives. Business leaders who answer this question will decide whether they want to reduce costs by eliminating their head count or generate more revenues by creating better products and services. Rather than replacing human workers, identic AI systems can help workers automate repetitive tasks, analyze complex data, and manage projects more effectively. As Brynjolfsson explains:

> If you have technology that imitates humans, it becomes a substitute, and it makes labor less valuable. But if you have a technology that augments humans, allows people to do new things, and humans are still

in the loop, then it makes human labor more valuable. We should have machines that can do wondrous things that we couldn't even imagine, especially in combination with humans.[25]

In industries such as law or education, AI systems may assist in generating legal briefs or creating lesson plans. However, lawyers and teachers must choose the direction of a case or the focus of a lesson. By reinforcing this dynamic, businesses can harness the potential of both artificial and human intelligence.

To benefit from both, businesses must adhere to key practices that promote agency:

1.  **Human-centered design**: AI designers must focus on human needs, not just efficiency. They must understand how AI can complement human workflows, giving users options rather than dictating actions. Businesses must foster collaborations between AI and human workers where people can override AI.
2.  **Empowerment through augmentation**: AI should automate routine tasks and augment human decision-making with insights. For example, in creative fields such as marketing or product design, AI agents can analyze trends, propose strategies, or generate content while human professionals can determine the creative direction and tailor strategies and content to specific audiences.
3.  **Training and upskilling**: Businesses can train employees to maximize the AI systems they use, so that they can oversee, modify, or challenge AI-driven output. User cohorts may come up with additional use cases or better ways to deploy AI within the organization.
4.  **Customization**: Employees should be allowed to tailor AI systems to their preferences so that they can control how they interact with these systems. They could set parameters for decision-making or adjust the frequency and type of AI recommendations.
5.  **Feedback loops**: Employers must implement feedback mechanisms

so that workers can correct or refine AI actions and output. Invite employee feedback on AI performance and adjust the system's decision-making processes to align better with human goals and values.

## Promoting Individual Autonomy

For individuals interacting with AI systems, safeguarding personal autonomy requires a proactive approach to managing AI:

1. **Understand AI's role**: Whether using an AI assistant to manage tasks or receiving AI-generated recommendations, individuals must recognize that AI is their tool, not their substitute.
2. **Customize AI tools**: Users must actively shape AI contributions rather than passively accept its outputs. Users should prioritize AI systems that they can personalize and align with their preferences and goals.
3. **Be cautious of overreliance**: While AI companions can assist individuals, particularly isolated ones, we must avoid overrelying on AI systems for our own needs or to care for other human beings. People need people, and people need physical and cognitive workouts. As AI systems grow more sophisticated, we can enhance personal agency by actively interpreting and validating AI-generated insights.

The key to preserving human agency is adopting a balanced approach to automation. "If we devalue human labor, and it leads to more of a concentration of wealth and power, then capital owners have all the power and capital is much more concentrated than labor," says Brynjolfsson. "Most of us would prefer a world with widely shared prosperity rather than highly concentrated wealth and inequality. So, for those reasons, I advocate for trying to keep humans in the loop. It creates more value, and it creates more widely shared prosperity."[26]

# ADAPTABILITY: THRIVING IN THE ERA OF IDENTIC AI

With AI systems reshaping industries, workflows, and daily life at a brisk pace, remaining flexible and continuously learning are critical to everyone's success. As Mark Cuban, entrepreneur and investor, recently advised, "Artificial intelligence, deep learning, machine learning—whatever you're doing, if you don't understand it—learn it. Because otherwise, you're going to be a dinosaur within three years."[27] That applies to companies and nongovernmental organizations, whose leaders must instill a culture of agility and resilience amid constant change.[28]

## The Adaptation Imperative

Adaptability in the age of identic AI refers to the organizational capacity to evolve with technological advancements, seamlessly integrating AI into business operations and individual workflows. For businesses, this means opening up to experimentation, rapidly iterating on new ideas, continuously assessing how AI can enhance operations, customer experiences, and decision-making processes, and shifting from rigid, hierarchical structures to more fluid, dynamic models that encourage innovation.

For individuals, it means developing the skills and mindset to collaborate with their identic agents, keeping up on new developments, and embracing rather than resisting change. Today's human workers will thrive when they learn to interact effectively with AI agents, automate routine tasks, and pivot to strategic and creative work.

## Cultivating a Culture of Adaptability

Nowhere is the culture of adaptability more evident than in AI-native firms—organizations designed around AI capabilities from the ground up. Unlike traditional corporate behemoths, these firms have no legacy structures and practices. Instead, they leverage AI to rework work.

Jared Spataro, Microsoft's chief marketing officer for AI at Work, and his team have extensively analyzed AI companies to understand their unique operational traits. First, their organizational structures are flatter. "Every manager can oversee a combination of agents and people, making traditional hierarchies less necessary," Spataro explains. "With broad access to agents and copilots, employees enjoy a level of support once reserved for top executives."[29]

Second, AI-native firms hire generalists over functional specialists. "Historically, organizations relied on specialized roles because expertise was scarce and expensive," Spataro notes. That was Adam Smith's classic economic principle: that increasing specialization fueled productivity. "But with AI agents, we now have expertise at our fingertips."[30]

Third, AI-native firms organize around outcomes not departments such as HR, sales, marketing, and operations. "These outcomes often span multiple agents, people, and data sources, breaking down the silos of departmental thinking," Spataro observes.[31]

In short, AI-native firms are flattening hierarchies, busting silos, democratizing support, and pooling generalists so that they can achieve their goals more quickly and efficiently.

Not all businesses can reorganize their structures and business models around outcomes. However, leaders can take these practical steps to modify their organizational DNA:

1. **Encourage AI experimentation**: Leaders should nurture an environment where employees can discuss, test, share, refine, and integrate AI tools into everyday operations. By making space for experimentation, businesses can identify the most effective applications of AI and stay at the cutting edge.

2. **Invest in lifelong learning**: As AI technologies evolve, the skills people need to work alongside them will also evolve. Businesses must offer ongoing learning opportunities, including partnerships

with educational institutions, online learning, and internal AI training and tutorials.

3. **Promote agility**: Companies must be agile enough to pivot when AI presents new opportunities or disrupts existing markets and business processes. Pivots could involve redesigning products, workflows, and jobs, and updating performance metrics accordingly. Agile project management, with iterative development and cross-functional teams, could help businesses stay nimble amid AI-driven change.

4. **Foster a growth mindset**: People with growth mindsets think they can cultivate their talents through good learning strategies, hard work, and feedback from others.[32] Leaders must encourage such a mindset throughout the organization so that employees experiment openly with AI in their work and share what they learned.

## Adaptability for Individuals in a Rapidly Evolving World

Individuals should take steps to stay relevant in the AI-driven economy. The most valuable human workers will be those who can harness AI to amplify their capabilities in tasks ranging from project management to creative brainstorming. Here are some action items:

1. **Develop AI literacy**: To work effectively with AI, understand how these systems operate, what they can do, and what they can't do yet. Try using AI tools and applications relevant to your field, such as those we've identified throughout this book.

2. **Stay curious and embrace change**: Employers will value those who adapt quickly to new tools and workflows in the AI-driven workplace. Stay curious about new AI developments and explore how they might affect your job, industry, or career path and take advantage of new opportunities as they arise.

3. **Leverage AI for personal growth**: Identic agents can also support personal growth, help individuals learn new skills, set and achieve goals, and even explore creative pursuits. Pick up these tools and stay competitive in the job market.

4. **Collaborate with AI**: Rather than viewing AI as a threat, see it as a collaborator that can assist with routine tasks, offer new perspectives, and help you focus on high-level strategic thinking. Integrate AI into your workflows for creativity, productivity, and innovation.

As AI continues to evolve and reshape industries, individuals and businesses must remain flexible, open to new ideas, and willing to embrace change. Those who can pivot quickly, learn continuously, and experiment with AI will thrive in the era of identic AI.

## FAIRNESS: MORE THAN ETIQUETTE

In the context of AI, fairness refers to the unbiased treatment of all individuals, regardless of background, identity, or demographic. When AI systems make decisions—whether in hiring, approving loans, assisting physicians, or serving customers—they must avoid perpetuating inequalities. Developers train the algorithms behind identic AI on massive datasets, which may reflect historical biases that affect real people in significant ways.

For example, in 2017, Google developed a tool, the Perspective API, to moderate comments automatically on platforms like YouTube.[33] With billions of videos and comments uploaded daily, YouTube needed AI to help human moderators assess comments for toxicity.

Functioning as a toxicity classifier, the Perspective API assigned each comment a score from 0 to 100 to indicate its potential toxicity. Google first launched a public demo so that users could experiment.

The demo was quite revealing. For example, when a woman entered the phrase "I am a woman," the tool assigned it a toxicity score of 40 percent. When she typed "I am a gay, Black woman," the score skyrocketed to 90 percent. In contrast, the phrase "I am a man" received a toxicity score of 20 percent. This discrimination stemmed from the training dataset. The dataset contained too many instances that cast identity terms like "woman" or "gay" in negative contexts and too few positive or neutral uses. Consequently, the algorithm associated these identity terms with toxicity by default.

Recognizing this problem, Google publicly addressed it. It retrained the model with more balanced data until the algorithm handled identity terms more fairly, and then it released a case study, a video on what went wrong, and posts on Medium in which it documented its steps to correct the algorithm. This example highlights a critical lesson: Developers must vet the balance and diversity they use to train their algorithms. If datasets overrepresent or underrepresent certain data elements, then AI systems can produce biased or unfair results.

Of course, comment moderation is the tip of the iceberg. In many circumstances, AI-driven decisions directly affect people's lives, from their access to education and employment to their eligibility for credit and healthcare. For businesses, fairness is not just a matter of ethical responsibility but a legal and reputational necessity. Fair AI can foster trust, create more scalable products and services, and protect businesses from costly regulatory actions or public backlash. Fairness for individuals means protecting their rights and opportunities from algorithmic discrimination.

## What Fairness Entails in Identic AI Applications

In identic AI, fairness goes beyond simply avoiding bias. X. Eyeé, CEO of Malo Santo, a responsible AI consulting firm, emphasizes that fairness is not a universal truth. "Fairness is based on subjective values that vary from person to person, making it almost impossible for an algorithm to be 100 percent fair," Eyeé explains.[34] Developing an AI model inherently involves trade-offs between different people's value systems.

To illustrate, Eyeé references the Perspective API's role in moderating hate speech. "The way the AI is programmed to recognize and respond to certain content reflects specific choices about what is considered acceptable or harmful," Eyeé says. "These choices are shaped by underlying values, and different groups may not always agree on them. This makes achieving perfect fairness extremely challenging."[35]

While algorithms are imperfect, researchers have found that, in some cases, AI can make decisions less biased than human ones, particularly where developers have thoughtfully designed and trained their AI models on balanced datasets.[36] We can break fairness into several components:

1.   **Inclusive design**: Fairness begins in the design phase. Consider the needs of different user groups and design the AI to serve all of them equally. Research how different user groups experience AI. For example, voice-recognition AI systems have historically performed better with some accents and dialects than others, often frustrating users whose speech patterns are less well understood. Make conscious design choices that anticipate and prevent such disparities.

2.   **Bias mitigation**: AI systems learn from data. If the data contains biased patterns, then AI can replicate and amplify those biases. Remove these biases before training models—or recall and diagnose algorithms with problematic outcomes. Use explainability techniques to understand what a model is doing. "After identifying the problematic pattern, you need to give the algorithm enough data to unlearn that pattern and relearn the correct one," explains Eyeé. "The challenge is to do this without disrupting all the other patterns it has learned—a risk known as *overfitting*. In essence, you're often left rebuilding the algorithm from scratch."[37]

3.   **Equitable access**: Strive to serve individuals from all walks of life. Promote equitable access to diagnostics and treatments in

healthcare, leaving no underserved population behind. For example, AI should not disadvantage students from lower-income backgrounds or marginalized communities.

## Practical Business Advice for Ensuring Fairness

For businesses deploying identic AI systems, the stakes are high. Deploying AI that is fair to all users isn't just good ethics; it's good business. As X. Eyeé explains, "I want my identic AI to perform well for all my customers—regardless of their identity. My goal is to maximize my potential to scale, because if my AI fails for people with darker skin tones, then I'll never be able to penetrate the market in the continent of Africa."[38]

According to Eyeé, businesses must ask whether all their customers, with infinite variations in needs, backgrounds, identities, and experiences, are having consistent experiences with their algorithms. "If my AI doesn't work, it shouldn't be biased in how it fails. It shouldn't fail specifically for women, or for people in certain zip codes, or for those with lower-quality internet or older phones," said Eyeé.[39]

Ultimately, fair AI unlocks growth, while unfair or biased AI decisions can damage reputations, increase legal liabilities, and erode customer trust. Here are some practical steps to maintain fairness:

1. **Bias audits**: Audit AI systems early to identify and correct biases before the AI causes harm, and audit regularly to catch bias that has crept into AI models through training data, model architecture, or decision-making algorithms.
2. **Diverse datasets**: Use datasets that represent diversity in the real world—all ages, genders, ethnicities, and socioeconomic backgrounds—and otherwise overlooked or underrepresented minority groups. Also mind data gaps that could skew results.
3. **Inclusive development teams**: Form development teams with varied backgrounds and experiences, encourage everyone to

speak up and express concerns about bias throughout the development process, and reward them for fast delivery of fair models.

4. **Fairness as a key performance indicator (KPI)**: Add fairness as a measurable outcome in AI performance by setting fairness goals alongside traditional metrics like accuracy or efficiency, defining fairness KPIs to measure how well the AI system performs across different demographic groups, and retraining or adjusting AI models to meet fairness benchmarks.

## The Role of Individuals in Safeguarding AI Fairness

Individuals must also watch for and ask about AI in use. Whether applying for a job, seeking a loan, or interacting with customer service agents, individuals may ask whether an organization is using AI in such processes. People can probe to determine whether AI is treating them equitably:

1. **Understand the role of AI**: Individuals have a right to be informed when AI is making decisions that affect their lives. Whether applying for a mortgage, a job, or using a mental health crisis service, people should know when AI is involved. So ask customer service organizations whether AI is informing decision-making related to your account, service, or experience and how the AI model determines recommendations, pricing, or moderation. Or ask your prospective employer if they are using AI systems in screening job applications. If so, what criteria do they evaluate?

2. **Question decision-making**: If individuals believe an AI system has mistreated them—whether rejected for a loan or passed over for a job—they should ask for an explanation. AI systems must be explainable, and companies are responsible for providing clear answers about how they process decisions. Self-advocacy is particularly important in finance, healthcare, and employment, where decisions have life-changing consequences.

3. **Advocate for accountability**: Advocate for accountability by asking companies for audits of their AI models for fairness, asking government representatives for laws that protect individuals from discriminatory AI, and supporting industry coalitions like Partnership on AI, which brings together companies, academia, and civil society to develop best practices and guidelines for responsible AI use.

4. **Participate in the conversation**: Contribute to discussions of how industries and society at large should govern AI. Check out the Responsible AI Institute and the Future of Life Institute to stay on top of advances in technology and proposals for AI guardrails. The Algorithmic Justice League (AJL), founded by Joy Buolamwini, also advocates for equitable algorithms and transparency in AI development.

Advocating for fairness in identic AI is not a onetime effort; it requires ongoing commitment and vigilance. As AI systems grow more sophisticated, the nature of bias will evolve, and businesses and individuals must stay proactive. Fairness must be a guiding principle in developing and deploying AI so that these powerful systems are tools for inclusion and equality rather than instruments of discrimination.

# ACCOUNTABILITY: GOVERNANCE AND RESPONSIBILITY FOR IDENTIC AI

Accountability is about establishing clear ownership, oversight, and responsibility for the actions and decisions of AI systems. For every autonomous action that an AI agent takes, a defined party—an individual or other legal entity—must be responsible for that outcome. Whether the person or party is legally accountable depends on context, the laws in place, and the outcomes of lawsuits. This accountability chain becomes increasingly important as AI systems take on more complex and autonomous tasks, from identifying illegal activities and assisting in surgical suites to shuttling passengers around cities and preventing midair collisions.

While AI agents give us unprecedented autonomy and decision-making power, their complexity raises significant concerns about who is ultimately accountable if an AI system makes an error, causes harm, or perpetuates bias. Having a clear answer to this question is vital for good governance and public trust in AI.

## Defining Accountability in Identic AI

In 2012, the State of Michigan's unemployment agency introduced the Michigan Integrated Data Automated System (MiDAS) to improve efficiency in detecting and preventing unemployment fraud. While the intention behind MiDAS was sound, the outcome proved catastrophic. Between 2013 and 2015, the system falsely flagged over forty thousand claimants for fraud, with an accuracy rate as low as 8 percent.[40] These erroneous accusations led to wage garnishments, intercepted tax refunds, and damaged credit scores for the affected individuals.

The lack of transparency in the algorithm's decision-making process made it difficult for individuals to contest the false accusations, exacerbating their plight. Moreover, the state's overconfidence in the system's infallibility delayed corrective actions, prolonging the suffering of those affected. The fallout was severe. In January 2024, the state agreed to a $20 million settlement to compensate those harmed financially and reputationally. The state also reevaluated its reliance on automated systems for decision-making.

Every day we see new examples of important decisions in people's lives made by artificial intelligence. But what happens when AI costs someone a job or unjustly labels an innocent person as a fraud? While automation can streamline operations and detect anomalies, the MiDAS case demonstrates that without proper checks and balances, such systems can cause more harm than good.

Crucially, developers or their employers must design accountability into AI systems from the outset and delineate responsibility across developing, testing, deploying, and monitoring AI systems. Without this, businesses and society face the risk of AI systems operating in harmful or counterproductive ways, with no mechanism for interrupting operations and correcting the code.

## Practical Steps for Business Accountability

Businesses seeking to integrate AI systems responsibly must make accountability a top priority. According to Ayanna Howard, they must give users the choice to opt in or out of AI systems. "AI makes mistakes, and we know it will never be one hundred percent accurate. But if I have the agency to choose to use AI, even with its imperfections, I will continue to engage with it, build trust, and comply with it," Howard explains.[41]

However, when organizations enforce the use of AI, persistent errors can erode trust. "If AI use is imposed on me and errors keep occurring, I will likely switch. I'll choose a different AI agent, a different platform, or even revert to a human alternative. Even though humans may be less efficient, I will trust them more in such cases," she adds.[42]

Howard's insights underscore a key aspect of AI accountability: the freedom to choose. Giving users autonomy not only builds confidence but empowers users to control their interactions with AI.

Businesses can also adopt additional strategies such as regularly auditing AI decisions, offering clear explanations for errors, and maintaining accessible feedback channels. These practices can help flag potential problems, mitigate risks, foster trust, and promote ethical deployment of identic AI:

1. **Define ownership at each stage of development**: AI systems go through multiple stages. At each stage, a team or organization must identify who is responsible for AI performance according to expectations and ethical guidelines. For instance, the AI

developers are accountable for training the system on diverse datasets to minimize bias. At the same time, the operations team with a client enterprise are accountable for monitoring and reporting on the system's performance.

2. **Establish clear lines of reporting**: Organizations must also establish clear lines of reporting for AI-related issues. If an AI system behaves unexpectedly, then someone or some team must be accountable for identifying root causes, taking corrective action, and preventing recurrences. Such processes likely require collaboration and a sense of urgency among IT, legal, and operations teams to diagnose problems, implement solutions, and communicate with users.

3. **Audit AI decisions regularly**: For themselves and their stakeholders, organizations would be wise to audit their AI systems as regularly as they audit their financials. How fair and accurate are our decision-making processes? Do they comply with relevant laws, ethical guidelines, and industry best practices? Businesses that conduct regular AI audits can show accountability to regulators, customers, and other stakeholders.

4. **Ensure compliance with AI regulations**: As AI technologies advance, governments are setting new regulations to govern their use. For example, the European Union's General Data Protection Regulation (GDPR) includes provisions for algorithmic transparency and accountability. More countries may introduce similar regulations. By staying ahead of these requirements, businesses can minimize their legal risks and engage positively with regulators.

## Holding Organizations Accountable: A Guide for Individuals

As AI systems influence more decisions that affect people's lives—from healthcare to jobs—people must know their rights and hold organizations accountable for AI-driven decisions. As when seeking fairness, the first step is identifying which organizations are using AI, for what, and how. The second step is monitoring outcomes. The third step is requesting an explanation if an AI system makes a decision or takes an action that affects you negatively. Once you have that information, consider the following actions:

1. **Advocate for redress**: Democracies that uphold the rule of law often have statutes, regulations, and enforcement agencies that protect the rights of employees, consumers, and citizens through, for example, some form of due process. It could be filing a dispute or an appeal through the appropriate channel of an organization, a government agency, or a court of law. If you feel that an AI system has made an unfair or biased decision against you, then advocate for redress through the channel closest to the decision. Terms of use/service and terms of employment often specify the channel and describe the process. Reaching out to consumer advocates, union representatives, nonprofit organizations such as those we've identified in this chapter, and law firms with AI practices can help, depending on your situation.

2. **Keep on top of AI rights and legislation**: Individuals should know their rights as the legal landscape for AI governance evolves. New regulations, such as the European Union AI Act, aim to create more safeguards around the use of AI in everyday life. By understanding these laws and the structures they put in place, individuals can hold businesses accountable for their AI systems.

Without robust accountability mechanisms, the output of the AI systems may harm individuals. While AI may be autonomous, the responsibility for its harms ultimately lies with those who created and deployed it.

## SAFETY: SECURING THE AGE OF IDENTIC AI

In 2023, the godfather of AI, Geoffrey Hinton, left Google to raise the alarm about AI: He said there was a 10 percent to 20 percent chance that superintelligent AI could bring about human extinction. The stakes are high as we forge ahead in developing AI.[43]

Hinton has explained that humans were on the verge of becoming the second-most intelligent species on the planet, with little control over AI's evolution. The problem was one of alignment: How can we align the goals and outcomes of AI systems with those of humanity so that AI benefits humankind? "What we want is some way of making sure that, even if they're smarter than us, they're going to do things that are beneficial for us," Hinton explained. He acknowledged the difficulty of achieving this, when bad actors may program robot soldiers to harm people. "And that seems very hard to me," he added.[44]

Although today's AI systems generally respond to user prompts and not their own objectives, Hinton warns that we must take seriously the risk of ambitious AI: "They may well develop the goal of taking control," Hinton said. "And if they do that, we're in trouble."[45]

AI becoming uncontrollable is not the only risk. Even if superintelligent AI does not drive human extinction, the list of its potential harms is long and expanding. AI-powered attacks on critical infrastructure, algorithmic bias that exacerbates social inequalities, and the widespread manipulation of truth all threaten human civilization.

"If AI models are much smarter than [we are], they'll be very good at manipulating us," Hinton said. "You won't realize what's going on. Even if they can't directly pull levers, they can certainly get us to pull levers. If you can manipulate people, you can invade a building in Washington without ever going there yourself."[46]

As we develop identic AI, the imperative is clear: We must create systems that are robust, transparent, and aligned with human values. As

Hinton warns, the future of humanity may depend on the decisions we make today.

## What Does Safety Mean in the Context of Identic AI?

Although we'd like to think that our AI agents will do our bidding, they might not.

Because they can learn, interact with their environment, and adapt in real time, they might behave unexpectedly—harmfully. How can we prevent unintended consequences and limit the potential attack vectors for malicious users?

## AI Safety Advice for Businesses

As organizations deploy identic AI agents to handle everything from customer service to complex decision-making, they must integrate safety measures into every phase of the AI life cycle—from design and development to deployment and ongoing monitoring—and counteract harmful actions and cut off the potential for additional harms.

1. **Secure development processes:** Developers of AI technologies must adopt secure coding practices and commit to building AI systems with safety and security in mind. They can implement rigorous testing protocols to identify vulnerabilities that could lead to unexpected behaviors or exploitation and incorporate security best practices such as encryption, secure data handling, and threat modeling to identify risks early in development. Additionally, companies must deploy AI systems to detect and respond to potential threats autonomously. For example, an AI agent responsible for cybersecurity should detect and mitigate potential breaches, such as isolating affected systems or initiating an audit

of the breach. Another common security tactic is to deploy honeypot simulations—fake data and restricted environments designed to lure bad actors into revealing their tactics. AI systems can then learn how to recognize and counteract adversarial attacks before they infiltrate real-world applications.

2. **Monitoring and oversight:** Businesses must set up systems that can track AI decisions, monitor interactions, and detect anomalies that may signal a malfunction or manipulation. Organizations can keep their AI systems aligned with safety standards by using tools that continuously audit AI activity. Suppose an AI agent decides to reject a healthcare claim. In that case, insurers should have checks in place to validate those decisions and intervene if necessary.

3. **Build in a kill switch:** A hard-coded kill switch or quarantine mode could allow humans to instantly disable or limit the capacity of AI systems if they behave unpredictably or dangerously—say in the event that autonomous weapons, financial trading bots, or attacks on critical infrastructure cause catastrophic damage. Of course, a sudden, full shutdown could disrupt essential services and harm innocent people when AI is operating critical services like healthcare or energy grids. But a quarantine or shutdown may be better than a catastrophic alternative.

4. **Support AI safety research:** Hinton and others recommend devoting more people to AI safety research. "Right now, there [are] 99 very smart people trying to make [AI] better and one very smart person trying to figure out how to stop it from taking over. And maybe you want to be more balanced," said Hinton.[47]

## Protecting Individuals from AI Risks

From managing finances to personal health, individuals must ensure that identic AI systems are safe and trustworthy.

1. **Be aware of misinformation:** AI systems can generate realistic videos, audio, and text that deceives users or sways their beliefs. For example, fake videos of public figures can spread propaganda, damage reputations, manipulate public opinion, and incite violence. These deepfakes can erode trust in trained journalists, independent news, and other institutions so that the public cannot distinguish the real from the fabricated. In extreme cases, malicious actors have used deepfakes—of, say, public figures making statements or taking actions that they never did—to influence elections, disrupt diplomatic relations, and undermine national security. The impact can be long lasting and difficult to reverse with the truth. Individuals must pay close attention to sources of information. Do those sources cross-check facts and verify authenticity before publishing?

2. **Safeguard mental and emotional well-being.** As AI systems train on their users' data, they may get better at influencing users' thoughts, behaviors, and emotional responses in subtle and pervasive ways. AI systems that engage with individuals—whether for financial advice, mental health support, or companionship—could reinforce and exploit vulnerabilities.

To mitigate such risk, individuals must cultivate their own digital literacy, set boundaries for AI interactions, and seek AI service providers who are committed to ethical, transparent, and psychologically safe practices. As identic AI evolves, safety measures must, too.

# A ROAD MAP FOR LEADERS

Individuals and businesses seeking to leverage AI must acknowledge the extraordinary potential and the extraordinary risk. Biases embedded in AI algorithms, existential threats to human agency, the loss of privacy, and the concentration of power and wealth in the hands of a few companies controlling AI infrastructure should wake everybody up.

We believe the seven principles we've outlined in this chapter will guide users and organizational leaders in navigating these challenges:

**Reliability:** Deliver consistent, accurate results.

**Transparency:** Make AI processes clear and understandable.

**Fairness:** Promote equitable outcomes for all users.

**Accountability:** Take responsibility for AI-driven actions as an organization and as an individual.

**Agency:** Preserve human autonomy; AI must enhance, rather than render irrelevant, our unique capabilities.

**Adaptability:** Stay agile and responsive to technological change.

**Safety:** Protect businesses and individuals from malicious threats, unintended harm, and rogue AI.

These principles are essential for navigating identic AI in a way that promotes human prosperity and ethical long-term success.

# Chapter 11

# MAKING THE BIG LEAP: SUCCESS IN THE AGE OF IDENTIC AI

Growing up, Joseph was passionate about basketball. He loved it and was good at it. By the end of seventh grade, Joseph was practicing and playing on junior varsity teams. Coaches from different high schools would let him scrimmage with their teams because they hoped he'd choose to attend their schools.

They didn't know what Joseph and his parents knew, though he didn't fully understand its significance: He was born with a curved spine. Over the years, they watched it closely. Joseph's parents tried everything to spare Joseph surgery. Eventually, the curvature became too severe, and the doctor said surgery was inevitable.

The ten-hour surgery itself was incredibly challenging. Afterward, Joseph had to learn to walk all over again. He couldn't play basketball at his previous level. But basketball was everything to Joseph. Everything he did was tied to it. He got good grades because his dad told him that, if he didn't, he couldn't play. Without basketball, he had to redefine himself—before he was thirteen years old. The loss took a toll on him. By the time he reached high school, Joseph felt lost and disconnected.

Then, in his sophomore year, Joseph's English teacher pulled him aside. "Snap out of it!" she told him. She saw a talent—a "superpower," as she put it—in Joseph. She suggested that he try speech and debate, because she believed he'd be great, maybe the best their school had ever seen.

Taking her advice was incredibly difficult, terrifying even. It meant changing his very identity. But Joseph gave speech and debate a shot. His first tournament went well, but he didn't win. In the second one, he started winning. In his junior and senior years, he was state champion, competing at the national level.

AI requires all of us to rethink what we're good at, how we compete, and where we add the most value as human beings. Redefining ourselves— letting go of what we really love doing—is scary. But to unlock our new potential, we must take that step and take it with caution, but we need not take it alone. Joseph had his parents, his teachers, and his speech and debate team members. If we persevere and trust our humanity, we will succeed as employees, students, creators, citizens, and individuals seeking companionship.

Let's review seven critical challenges for each of us, our organizations, and our communities as we chart a course into this new age.

## REDEFINING PRODUCTIVITY, LEADERSHIP, AND SUCCESS AT WORK

In chapter 4, we described how identic AI will reshape work by enhancing individual productivity and automating many essential business functions. AI will serve as powerful personal assistants, handling routine tasks, analyzing data, and making real-time recommendations so that human beings can focus more on strategic, creative, and high-value activities. If we integrate AI into daily tasks, then we will free ourselves to respond to the all-new demands of AI-powered industries.

Companies will also unleash swarms of autonomous AI agents across business functions, such as logistics, finance, customer service, and supply chain management. On a recent earnings call, Salesforce CEO Marc Benioff called it the "rise of digital labor" and argued that businesses are "at the edge of a revolutionary transformation."[1] In September 2024, Salesforce launched Agentforce, an enterprise solution with out-of-the-box agents that users can easily customize and deploy wherever needed. "They're working 24/7 to analyze data, make decisions, take action, and we can all start to picture this enterprise managing millions of customer interactions daily as agents seamlessly resolve issues, process transactions, anticipate customer needs, and free up humans to focus on the strategic initiatives and build meaningful relationships," said Benioff.[2]

Industry players like Sam Altman are predicting that companies with just ten employees augmented by AI agents will achieve what once required a workforce of thousands at unprecedented scalability and operational efficiency. As a case in point, Wojtek Gudaszewski, chief operating officer of the online insurance company Nsure, said that, with AI agents, the company can grow the business with only 20 percent of the workforce it would otherwise need.[3]

## Thriving in the AI-Driven Workplace

To prosper with AI at work, workers must view identic AI as a tool that can enhance their potential rather than threaten their jobs. According to Susan Doniz, CIO of Boeing, the AI experience at big manufacturing companies has been less about reducing head count and more about restructuring work so that human beings do more and enjoy it more. "Instead of taking six hours of tedious work to write a work specification, they might take an hour because their AI provides good suggestions," said Doniz.[4]

Naturally, some see autonomous AI as a competitor where it can automate certain tasks and business functions. However, like the transition from manual labor to managing machines during the Industrial Revolution, workers must adapt to working with AI and honing their uniquely human skills that machines can't easily replicate. Companies will value emotional intelligence, critical thinking, creativity, and ethical decision-making more because these skills complement rather than compete with AI-driven processes. As Doniz put it, "Organizations are going to work differently. Instead of reducing head count we will focus on reducing the mundane work and redeploying our employees to do things that add more value to our company."[5]

Workers must also grow as people and develop their careers. To thrive in an AI workplace, employees must learn to use AI-driven analytics; interact, collaborate with, and even treat AI copilots as partners in their tasks. Jared Spataro, Microsoft's chief marketing officer for AI at Work explains how the employee relationship with identic AI could evolve. "Today, copilots are very reactive," says Spataro. "You have to ask them a question, often knowing the right question to ask and what the copilot is capable of. In the future, copilots will be very proactive and focused on helping you accomplish the results that matter to you—what you might call 'outcomes that matter.' We foresee a day when no one will want to do a job without a copilot."[6]

For instance, companies in the marketing and advertising industry are recruiting AI marketing specialists to interpret AI-generated insights, set up automated campaigns, and weigh ethical considerations in AI-driven ad targeting.[7] The shift is similar across healthcare, finance, and logistics, where companies are creating new hybrid roles that combine human judgment with AI's data-processing abilities. Of course, when hiring new employees, recruiters will consider how well candidates have curated their agents and their effectiveness as a package. In the start-up world, entrepreneurs with superpowers will tend to do better than without them.

Next-generation skills will be vital in this transition. Workers need proficiency in querying AI systems, adjusting prompts and settings to achieve specific goals, and managing the automation of routine workflows. Critical thinking will be equally important. Workers must learn to curate and interpret AI outputs effectively. Accenture's chief technology and innovation officer Paul Daugherty says the reskilling challenge is massive. "Not only do you need digital skills," said Daugherty, "you need the sophistication to interact with digital agents in an intelligent fashion."[8] For example, a business analyst would want to understand how AI arrives at conclusions, critically evaluate its recommendations, and collaborate with AI to align strategies with key business goals. Customer service agents would want to repurpose their skills so that they could complement digital agents.

Lifelong learning is no longer a slogan, it is becoming essential, since AI is always learning. Even retirees may find themselves interacting with AI around their retirement benefits or for that matter anything else. As with the commercialization of the internet in the 1990s, the future workplace will favor those who adapt quickly to digital technologies, can pivot to new roles, and think critically about how their skill sets complement AI.

## Rethinking Leadership for the Era of Identic AI

Chief executives must also shift their mindset to lead in radically different business environments. Beyond incrementally improving businesses, AI is fundamentally transforming how businesses operate.

First, it accelerates the pace of operations. CEOs must move away from periodic business planning to dynamic planning, where AI continuously analyzes vast amounts of data to uncover risks and opportunities that human leaders might not see immediately and to act or to recommend acting on those risks and opportunities.

Second, CEOs and executive committees must review their control frameworks to determine which decision rights and resources AI should have across the organization. In a 2024 Accenture workforce survey, nearly all workers (94 percent) said they were ready to learn AI skills, but just 5 percent of organizations were providing training at scale. Meanwhile, 60 percent of workers said they were worrying about AI eliminating their jobs; and two-thirds of executives said they had neither the technological nor the change leadership expertise to leverage identic AI fully in transforming their organizations.[9]

This gap in C-suite readiness should startle boards and shareholders. If your organization isn't already reskilling and upskilling employees to work alongside AI systems, and if you're not cultivating trust in the teams responsible for executing AI-supported strategies, then you're already a laggard. By decentralizing decision rights, CEOs can keep their companies agile, innovative, and responsive to market demands more quickly than competitors.

It's unclear how fast radical new business architectures will grow, but evidence from DAOs and other new networked business structures foretell far-reaching change. In the internet age, savvy executives took advantage of outsourcing, and business models became a source of competitive differentiation. The AI age will amplify such opportunities exponentially "to the power of two."

Finally, leaders must navigate uncertainty. AI's full potential is largely unknown and unknowable right now. While we can predict many AI advancements, others will emerge on their own, and we may comprehend them in hindsight. CEOs must see themselves less as leaders who map out the future and more as stewards who guide their organizations through ambiguity.

Leaders must combine their intuition with identic AI capabilities to ask the right questions of AI. Consider one of the influences in Joseph's career, Marvel Comics. For years, this great comic book company focused on "How do we license more of our TV and film rights?" This question guided its strategy for decades. But somewhere along the line, someone asked, "What if we make our own TV series and films?"

This story reminds us of the power of human imagination. Human insight helps define the broader purpose, aligning AI with long-term vision and values. When humans set nuanced, visionary goals, identic AI can complement that vision by executing complex strategies and making adaptive decisions. Through this partnership of human foresight and AI execution, organizations may achieve meaningful outcomes that resonate commercially and culturally.

## LIFELONG LEARNING: A NEW APPROACH TO KNOWLEDGE

In chapter 5, we described how identic AI could revolutionize education by personalizing learning experiences. We also introduced a historic new challenge: learning to collaborate with your digital self—and its near-limitless access to the world's recorded information—serving as your auxiliary memory, knowledge repository, and personal tutor.

AI companions that adapt to individual learning styles and paces will tutor students of all ages, from toddlers to the never retired. These AI companions will monitor progress, give immediate feedback, and guide learners through complex subjects. Meanwhile, human teachers will offload basic instruction and many administrative tasks to AI so that they can focus on the social, emotional, and ethical aspects of acquiring and applying new skills and knowledge. Imagine if adults—at home, at school, and at big tech companies—had known to do that at the advent of social media.

As individuals enter the workplace, AI companions can serve as career coaches, identifying gaps in individuals' knowledge, suggesting courses for professional development, and curating career networks and resources. In essence, identic AI will cultivate an environment for continuously learning and upskilling so that individuals stay open to the new and relevant in ever-changing job markets. Beyond traditional workplaces and academic environments, AI companions could help individuals learn new languages, practical skills, or hobbies and otherwise guide them on adventures around the world, in person or virtually, to the ends of their lives.

## Forging a Personalized Learning Journey

If the schools you attended—the classroom environments, the forms of instruction, or the contents of courses—never worked for you or aren't working for you now, then we've great news: You can create your own. The School of You, mascot optional. Around the world, parents are giving their children this option, with dynamic homeschooling curricula and communities.[10]

But gone are the days when individuals learned only in the first few stages of life. Lifelong learning is essential. Our careers will evolve multiple times over our lifetime. So why not embrace continuous learning and make it our own, not as a chore but as a creative and dynamic process that evolves as we grow and as our interests change? AI companions can point us to relevant courses and content and help us stay competitive in an ever-changing job market.

Some of us may discover that, after taking charge of our own education, we'd rather launch our own company, too. Our AI companions can guide us in achieving that goal, mapping our learning to the growth of our business propositions, our company, and our leadership. To succeed, young people must see learning as a constant, evolving, and exciting part of their lives.

Also essential is critical thinking. No matter how advanced, AI systems work only with the data developers give them, and data can be incomplete or biased. They are tools, not oracles. Therefore, young people must be able to question, interpret, and apply AI-generated content and information critically, even skeptically. By asking the right questions and scrutinizing the answers, young learners actively participate in their education and learn to apply knowledge in meaningful, thoughtful ways.

Finally, we must embrace the power of interdisciplinary learning. One area of expertise will no longer suffice for long-term success. Young learners must evolve their portfolio of skills and knowledge domains continuously to stay relevant as AI reshapes industries. Beyond technical skills, workers need interdisciplinary knowledge in such fields as AI, design, data science, and social sciences to tackle complex problems. Young people should develop a T-shaped skill set, where they have depth in a particular area and breadth across multiple disciplines. Identic AI companions can guide them through new fields, identifying connections between them. With this interdisciplinary mindset, young learners will prepare themselves for work environments where skills must evolve as quickly as technologies.

In short, identic AI offers all of us unprecedented opportunities to pursue lifelong learning. To succeed at work, we must continuously learn, think critically, and develop skills and knowledge across disciplines.

## Redefining the Role of Educators

The rapidly changing educational landscape demands significant changes from teachers. The traditional model of education places educators at the center as the primary source of knowledge. However, with AI companions providing instant access to vast information, educators can shift from knowledge providers to learning facilitators.

This shift reorients the roles of educators significantly. Rather than delivering standardized course content, educators will guide students in personalizing learning pathways, navigating AI tools, asking the right questions, analyzing information, and thinking independently. Educators can help students cultivate the skills they'll need to leverage AI for lifelong learning.

Like students, educators must adopt a mindset of continuous learning and professional development, staying up to date on AI tools and educational technologies in their teaching practices. Mary Lacity, David D. Glass Chair and distinguished professor at the University of Arkansas, says educators must not only use AI proficiently but also know how to evaluate and adapt AI-generated content to their students' needs. "If I'm such a bad teacher that I'm asking something that an AI can get 100 percent on, then that's on me, right?" said Lacity. "It's up to me to design an assessment that is inspiring to the students, challenging the students."[11] Lacity calls on educators to blend pedagogy with technology to create more engaging and adaptive learning experiences. And, like their students, she says educators must constantly learn, unlearn, and relearn to be successful.

The final shift speaks to the essential task of forming and maintaining the human connection in education. One of AI's key advantages is its ability to personalize learning experiences. AI can assess students' strengths, weaknesses, and learning preferences, providing tailored educational content that optimizes learning outcomes. But AI cannot replicate the empathy, emotional support, and mentorship that human educators provide. Educators will need to preserve humanity in education so that students feel supported, valued, and meaningfully engaged with each other.

Prioritizing the social dimension of the school experience is critical. Parents and teachers can emphasize group activities such as sports, clubs, and collaborative assignments, all vital to cultivating essential social skills that students will need later in professional environments. These activities foster collaboration, communication, and empathy that AI systems cannot easily replicate.

## NAVIGATING THE CREATIVE FRONTIER

Chapter 6 documents how autonomous AI agents are reshaping creative and cultural industries, such as art, music, and journalism. For the first time in history, creators now have full-time copilots in artistic processes, helping to generate visual art, compose music, and write news articles. High-quality AI tools make creativity and cultural production more accessible to amateurs and nonprofessionals. Yet, these advancements raise ethical questions. How authentic is AI-generated work? Does AI diminish the value of human creativity? How will registering and managing intellectual property rights change when creators and inventors use AI in their creative processes?

AI also significantly affects journalism, with tools for content creation, data-driven reporting, and efficiency. However, overreliance on AI in newsrooms raises the risk of misinformation, biased content, and the erosion of investigative journalism. In short, on this new creative frontier, the pioneers must carefully manage their usage so that AI-driven innovation in arts and media does not jeopardize human creativity, critical thinking, and societal trust.

## Adapting to AI-Curated Creativity and Media

In the past, we could easily distinguish between human-generated and machine-generated content—but no longer. Intelligent algorithms and generative AI will reshape how we consume information and entertainment. More than before, consumers need critical thinking and digital literacy to identify AI-generated content and combat the spread of misinformation. Today's AI technologies can create near-perfect simulations that deceive even the most discerning viewer. AI-generated videos that convincingly alter or fabricate people's appearances and voices threaten public trust in media. So, too, does the proliferation of AI bots masquerading as genuine people on social media.

By some estimates, bots now account for nearly half of all internet traffic globally and are ubiquitous on social platforms like Facebook and X.[12] AI developers program these bots to engage in conversations, spread misinformation, or simulate public support for particular ideas, all while appearing to be genuine users. Bots can distort public perception, manipulate discourse, and contribute to the spread of false narratives, further undermining trust in online platforms. Together, these AI-generated manipulations challenge the integrity of media and the transparency of public communication.

Consumers must learn to question the veracity of the media they engage with. By critically analyzing the sources of information and cross-referencing them with trusted outlets, individuals can safeguard themselves from fake news or manipulated media. Just as fact-checking has become an essential part of consuming online content, so, too, will verifying the authenticity of audio-visual media. Furthermore, consumers should familiarize themselves with emerging AI tools that detect AI bots and deepfakes.

Consumers must also demand transparency and think critically about the hyperpersonalization of digital media. As identic AI becomes more entrenched in entertainment—whether through algorithmic content curation on streaming platforms like Netflix and Spotify or through an AI companion's personalized recommendations—consumers must proactively hold these entities accountable for how they deploy AI to solicit online engagement.

Next, consumers must watch how algorithms curate their experiences. Our digital selves will shape our cultural consumption, from personalized playlists and social feeds to news and movie suggestions. While AI can enhance discovery and tailor entertainment to individual preferences, it can also narrow consumers' exposure to diverse ideas, forming filter bubbles that limit creativity and open-mindedness. As consumers, we must ask: Does the content my AI agent is feeding me reflect my genuine interests, or is the algorithm seeking to maximize engagement? Demanding transparency from content providers, encouraging ethical AI use, and supporting platforms that celebrate diversity can help mitigate the risks of overreliance on AI curation.

## Augmenting Human Creativity in the Age of Identic AI

Artists, journalists, musicians, and other content creators must also shift their mindsets to collaborate with AI systems and integrate new technologies without compromising their unique human talents.

First, creators must view AI as a creative partner that can enhance and expand their work, not replace their artistic vision. Creative illustrator Tomer Hanuka says he has spent years cultivating his style, brand, and artistic identity—the characteristics and capabilities that distinguish him in an increasingly AI-saturated marketplace. "I call it my cultural capital," said Hanuka. "It's name recognition paired with a certain visual approach that gives us power and a place in the conversation."[13]

While AI is no substitute for artistic vision, AI models like OpenAI's DALL-E and GPT-4 have demonstrated their power in generating new ideas, producing artwork, and writing stories. Creators like Mario Klinge-mann and Sougwen Chung are leveraging AI capabilities to push their own creative boundaries.

Musicians can also use AI tools to compose melodies or experiment with new soundscapes. Writers can use AI to help brainstorm plot ideas or generate character dialogue. Visual artists can incorporate AI-generated patterns into their work or create interactive art that responds to real-time data. By embracing AI as a creative tool, creators can stay at the forefront of innovation while retaining their emotional and cultural insights and their distinct approach to the craft.

In the second critical shift in mindset, creators must adapt traditional workflows and embrace interdisciplinary knowledge to stay competitive in industries where AI can accelerate content production. That means mastering AI tools for generating content, managing audience interactions, and monetizing creations. Journalists, for instance, might rely on AI to sort quickly through huge amounts of data. Visual artists might collaborate with AI to design immersive digital experiences.

Interdisciplinary thinking and capabilities will be vital. The convergence of art, technology, and science is more pronounced than ever, and the most successful creators will be those who can navigate across disciplines. For instance, AI-generated art often incorporates elements of machine learning, mathematics, and computer science alongside traditional creative practices. Journalists might collaborate with data scientists to uncover hidden patterns in massive datasets or use AI to augment investigative reporting.

With interdisciplinary knowledge, creators can innovate in ways that previous generations could not have imagined. Artists who understand the technologies' underlying AI systems will know how to push creative boundaries. Journalists and newsrooms working with AI tools will gain an edge over those not using them. Creators who stay curious, learn new skills, and blend creativity with technology will thrive.

The final shift involves managing intellectual property—art, music, or written content—that AI created or replicated without permission. As the lines between human and AI-generated content blur, lawmakers and IP offices must ask and answer questions about copyright, ownership, and authorship. Author Cory Doctorow warns that "neither AI companies nor entertainment companies will pay creative workers if they don't have to. Rather, they are calculating that they have so much market power that they can sell whatever slop the AI makes, and pay less for the AI license than they would for a human artist's work."[14] And many users may buy it. Doctorow cautions creators to become adept at navigating the legal aspects of intellectual property so that they can adequately protect their work.

One significant challenge arises when developers train AI systems on existing works of art or music without licensing the rights to do so, potentially generating outputs that closely resemble or even infringe again on the rights of existing intellectual property rights holders. AI developers and creators alike should familiarize themselves with emerging policies and laws governing AI-generated content and actively shape policies and statutes that protect their rights while promoting fair use and innovation in the jurisdictions where they create and sell their work. As AI contributes more to the creative process, creators must vigilantly defend their right to fair compensation for their original works, just as the Hollywood writers and actors did in their groundbreaking 2023 strikes.

## CIVIC ENGAGEMENT: EMPOWERING CITIZENS IN AN AI-DRIVEN SOCIETY

Chapter 7 describes an urban landscape where AI agents and autonomous systems optimize infrastructure, improve public service, and publish real-time data to inform civic discourse and decision-making. AI companions will streamline public services so that citizens can navigate city bureaucracy and access services. At the same time, autonomous mobility solutions will reduce congestion and help lead to cleaner, safer, and more efficient urban environments.

Most critically, AI can enhance civic participation, if we use it responsibly. Identic AI could aggregate citizens' input to gauge public sentiment on local issues. Individuals could use AI to engage more deeply in local governance. Together, these advances could make smart cities more inclusive and responsive to citizens' needs.

## Moving from Echo Chambers to Civil Cooperation

AI could empower citizens of urban centers to engage in local politics. AI-driven platforms can analyze enormous amounts of data on public opinion, policy impact, and community needs, giving individuals information to participate effectively in civic life. But citizens must shift from fulfilling political decisions to actively shaping them and governing.

Much as the transition from print to digital and social media reshaped how citizens engaged with the news, identic AI will transform how people interact with political systems. Citizens must adopt a mindset of vigilance and critical thinking, recognizing the power and limitations of AI in shaping public discourse and policy. They must advocate for transparency and accountability so that governments use fair, unbiased AI systems to serve the public good. Finally, because civic participation demands collaboration among diverse stakeholders, citizens must adopt a mindset of inclusivity, leveraging collective intelligence and engaging in open dialogue to reach more dynamic, equitable outcomes for their communities.

According to Andrew Rasiej, co-founder of Civic Hall and the Personal Democracy Forum, citizens needn't wait for governments to solve their problems when their AI copilots could help them self-organize. "We need a political culture where we celebrate civic engagement as a lifelong activity, one where participation and problem-solving aren't tied solely to voting for political representatives," said Rasiej. "Instead, our agents could essentially assist us in leveraging our interests and skills to improve our local communities, to support our neighbors, and for our neighbors to support us."[15]

A second big shift calls for proficiency with the tools of citizen engagement. Citizens must understand how to interpret and use civic data responsibly, engage with smart city dashboards, access real-time data on public services, and contribute to community feedback platforms. To do so, citizens must also understand how AI systems work, how they collect data, and how politicians and bureaucrats use data to inform their decisions. Citizens must also learn to use platforms for engagement and interactive decision-making, like crowdsourced urban planning initiatives or AI-assisted citizen juries. By more deeply understanding these tools, citizens can participate effectively in policy discussions, advocate for their needs, and hold local governments accountable.

Finally, we must address the political polarization of our political discourse. In today's digital world, people increasingly consume news and information that aligns with their beliefs, which algorithms often reinforce by prioritizing content likely to engage them. This phenomenon fuels polarization and forms echo chambers where people never hear, let alone consider, other perspectives.

Participatory democracy works only if citizens embrace a diversity of thought and actively seek out varying viewpoints to foster richer, more balanced discussions and civic engagement in local politics. Identic AI can help if we program the algorithms to deliver broader perspectives and even mediate dialogue between individuals with different backgrounds and political beliefs.

Ultimately, the most challenging shift for citizens will be building bridges between different ideological groups. Such collaboration is essential for solving complex challenges. Identic AI could offer balanced, data-driven insights that encourage fact-based debate and discourage misinformation. However, designers must consciously develop purpose-built identic AI systems that foster deliberation, cooperation, and empathy among diverse community members.

## Building Dynamic, Inclusive, and Participatory Cities

For elected officials and public administrators, identic AI presents opportunities and challenges in making cities more dynamic, inclusive, and participatory. The first big shift is prioritizing transparency, openness, and innovation. Government agencies often lag years if not decades behind everyone else in adopting and deploying digital technologies. "Everyone talks about how AI can help government deliver services better, but that's like talking about the penthouse before you've even built a foundation for the building itself," said Rasiej. "City after city, government after government, has failed to take full advantage of innovative possibilities that digital technology enables."[16]

A good place for public officials to start with identic AI is to commit to open data and make the data driving city decisions accessible to the public. Leaders should also explain how AI algorithms and public input shape decisions. Without transparency, the public's concerns about surveillance, privacy, or bias may outweigh attempts to foster trust and encourage public participation.

The second shift is fostering participatory leadership. With AI systems capable of gathering and processing vast public feedback, officials must jettison secrecy and backroom dealmaking. Instead, their use of AI tools should transform policymaking into an ongoing dialogue between government and the public. Rasiej suggests investing massively in professional development for public servants just as the private sector is training employees to harness AI systems.

A big culture shift is also in order. The top-down governance model must give way to a collaborative, grassroots model where elected officials use identic AI to hear voices from diverse communities. Through shared decision-making, leaders can ensure that tomorrow's cities reflect the needs and values of all their inhabitants.

The third and final mindset shift is adopting a proactive approach to ethical governance. Indeed, AI developers and public officials must commit to ethics and inclusivity as the principles of AI initiatives, not as afterthoughts. Most AI developers have set up voluntary frameworks for AI responsibility to govern their efforts. But taking a back seat to the technology giants is not a long-term solution. "The horse is out of the barn," says Rasiej, and "governments have shown up late again. They are thinking about it only in terms of fear, and designing regulations to address yesterday's challenges."

Like Rasiej, we believe elected representatives and public administrators must lead with foresight and align the deployment of AI in urban settings and other domains with democratic ideals and ethical standards. Together, technologists, ethicists, and community leaders can contribute to robust frameworks that address AI biases, ensure equitable access to AI-powered services, and safeguard data privacy. Working together is the only way to make these systems promote fairness and accountability.

## HEALTH AND WELL-BEING: THE SHIFT TOWARD PREVENTIVE, PERSONALIZED CARE

Chapter 8 describes a radical new model of personalized healthcare and wellness management enabled by a digital doctor who combines recorded medical knowledge with real-time knowledge about our bodies. AI health companions will track day-to-day vital signs and project long-term health trends. They will also provide real-time, data-driven advice based on an individual's unique biology and lifestyle. In doing so, identic AI vastly increases the supply of high-quality health advice, alleviating some of the constraints of today's overburdened healthcare infrastructure. It also replaces the reactive healthcare model—where we seek treatment only when something goes wrong—with a proactive, preventive model, in which AI anticipates health issues before they manifest and helps individuals make better decisions for long-term wellness. For example, AI companions will analyze real-time data from wearable devices and generate personalized diet, exercise, and medication management recommendations. This individualized approach extends to managing chronic diseases, like diabetes or hypertension.

At the same time, identic AI is revolutionizing medical research. Autonomous AI agents can analyze data faster than humans while identifying patterns and insights that scientist might not notice. DeepMind's Alpha-Fold, for example, predicts protein structures with remarkable accuracy to help scientists understand diseases at a molecular level. Capabilities like these will accelerate medical research, drug development, and guidance for preventing disease. As identic AI evolves, it will significantly move us toward more personalized, proactive, and efficient healthcare systems.

## Participating in the Personalized Healthcare Paradigm

Under the new personalized health and wellness paradigm, individuals must rethink their relationship with healthcare. Rather than rely solely on doctors and healthcare institutions, individuals will actively manage their health with AI companions' guidance. In this first big shift, individuals must move from passive recipients of generic healthcare services to active participants in personalized health management.

Shifting starts with a proactive mindset to managing your health. With identic AI, you can foresee health issues before symptoms appear. For instance, your AI health companion might detect changes in blood pressure, glucose, or cortisol levels early on, prompting you to change your lifestyle or call for medical interventions. "That's really the superpower of AI," says Arianna Huffington, the founder and CEO of Thrive Global, a well-being and behavioral change platform. "It's the hyper-personalization of the recommendations, nudges, and micro-steps enabled by an AI copilot that knows everything about you because it has collected billions and potentially trillions of data points."[17] Of course, embracing this new paradigm also means understanding that AI can help optimize health but not replace a trust-based relationship with local healthcare providers.

A second big shift is enhancing one's digital literacy to navigate AI-driven health tools effectively. Many individuals are already familiar with today's digital health apps. Identic AI seamlessly integrates their data daily to inform health management and decision-making. AI-powered tools, such as wearable devices, track everything from heart rates to sleep patterns in real time. They analyze patterns in individuals' health metrics to create wellness plans that adapt to the unique rhythms of everyday life. As Huffington put it, "I walk into the room and the AI knows I'm stressed, the music changes, my reset video comes up on the screen, and my AI is measuring my blood sugar and my vitamin levels and telling the robot in the kitchen what meal to prepare."[18]

Individuals must understand how these systems work, how to interpret AI-generated health insights, and how to adjust settings to personal health goals. Of course, individuals must also leverage health data to make lifestyle adjustments and more informed decisions about diet, exercise, and mental well-being. In short, the more individuals understand their health and wellness, the better able they will be to improve their health outcomes.

The final shift in mindset involves adopting a long-term, holistic perspective on health. Today's digital health solutions are well suited for short-term improvements, like helping users set up a workout routine or track nutrition. Identic AI fosters sustainable health and wellness over time. Of course, receiving the advice is only half the equation; acting on it is quite another. Individuals must think beyond immediate fitness goals to long-term health objectives, such as managing chronic conditions or preventing disease.

AI companions can help individuals by sending reminders, offering encouragement, and adjusting health goals as life circumstances change.

But this demands a mindset that views health as a journey rather than a destination. With AI's potential for early detection, preventative care, and personalized medicine, individuals who adopt a long-term approach will more likely enjoy healthier lives for years to come.

## Reinventing the Practice of Medicine

Doctors and other medical practitioners train for years to obtain mastery in their chosen fields, whether general medicine or specializations like cardiology and pediatrics. Many consult each other on puzzling cases and high-risk procedures. Still, their first big shift in mindset is the most significant: They must see themselves as trusted partners of AI when making medical decisions.

Unlike doctors, AI agents can quickly process sizable amounts of data: patient histories, genetic information, and real-time health metrics. Practitioners must learn to incorporate these AI-generated insights into their decision-making to pinpoint conditions or identify treatment options they might not have considered or learned about yet. In doing so, healthcare practitioners can deliver more accurate diagnoses, personalize treatment plans, and improve patient outcomes.

They can also save time. Dr. Peter McCaffrey, the University of Texas Medical Branch's director of pathology informatics and chief AI officer, said AI copilots can do medical interpretations of patient tissue or cell samples in seconds, where human beings took fifteen to twenty minutes per case. "It's a huge efficiency gain," said McCaffrey, "and it allows us, as pathologists, to do them more often and for more people, so we think that's a general public good for patients."[19] In short, doctors should see AI not as a competitor but as a team member that can enhance their diagnostic and treatment capabilities.

As the second big shift, upskilling is a pronounced imperative. Healthcare professionals generally embrace lifelong learning—indeed, many jurisdictions have long required continuing education credits for license renewal—and mastering the latest technologies and methodologies. AI in healthcare is evolving rapidly, and practitioners must update their skills continually, not simply to remain relevant but to improve more lives. Whether learning about AI-driven diagnostic systems, robotic surgery assistants, or AI models for drug discovery, healthcare professionals must continuously learn in this ever-changing domain.

Finally, for medical researchers, identic AI fundamentally shifts how they approach their work, particularly their research methodologies and data analysis. This shift moves them from traditional, labor-intensive research processes to AI-accelerated discovery and innovation. AI systems can now analyze massive complex datasets, identify patterns and potential drug candidates, simulate biological processes to predict treatment outcomes, and generate hypotheses faster than human researchers could alone. Researchers must rethink their unique role in advancing medical knowledge and develop the skills to interpret AI-driven insights, adjust research parameters based on AI findings, and collaborate with AI systems to drive faster, more accurate scientific breakthroughs.

Medical researchers must also cultivate an openness to interdisciplinary collaboration, as identic AI will frequently pull data and insights from various fields—biology, chemistry, computer science, and engineering. This cross-pollination can lead to innovations previously impossible in research silos. With these AI-driven methodologies, researchers will push the boundaries of medicine and contribute to advances in personalized treatments and precision healthcare.

## COMPANIONSHIP AND BELONGING: BUILDING COMMUNITIES IN A DIGITAL WORLD

In Chapter 9, we explain how the age of identic AI is transforming communities and interpersonal relationships as we integrate AI companions into our lives to fill roles ranging from mental health support and elder care to companionship. For those experiencing loneliness or social isolation, AI companions offer a synthetic emotional bond that can mimic authentic human companionship. In elder care, AI companions can engage in conversations, offer gentle reminders, and track health metrics, assisting caregivers and offering family members peace of mind.

Beyond individual relationships, identic AI has enormous potential to transform communities by breaking down barriers of distance, language, and accessibility and linking like-minded peers across cultural and generational divides. AI companions could help form dynamic, purpose-driven communities, catalyze collective action on a global scale, and redefine how we pursue our personal interests and advocate for causes we care about.

## Seeking Companionship in the Digital Age

Relying on your identic agent as a constant companion does require vigilance, particularly in dating, romantic relationships, and friendships. AI companions are superficially empathetic and responsive. While they may help users feel less lonely or even provide genuine companionship, they lack the deeper, reciprocal emotional intelligence and genuine vulnerability that human relationships require. As German futurist Gerd Leonhard put it, "There's happiness and then there's hedonism. Having a relationship with a hologram is hedonism, it's not real. An AI can simulate that friendship, but it can't replicate authentic human understanding, including the nuances and subtleties between lines of code."[20] Thus, as human and machine interactions blur, we must not replace or diminish the value of our human relationships with AI-driven interactions.

Second, individuals must be vigilant when granting AI a significant role in matching partners based on highly personalized and data-driven insights. Our AI companions can predict compatibility based on behavior, preferences, and communication patterns. While they can streamline finding partners, individuals should not allow algorithms to dictate their romantic choices. Users should cultivate self-awareness and prioritize meaningful human engagement over algorithmic efficiency. By all means, use your AI companion to curate a list of ideal partners and even pre-screen them by interacting with their agents. But balance AI-driven as-sistance with personal intuition and emotional intelligence to maximize your chances of developing authentic, fulfilling relationships.

Finally, older adults and all individuals seeking mental health support could supplement one-on-one and group therapies with AI companions and mental health apps like Woebot and Wysa. These apps already provide users with 24/7 support through conversational AI so that users can man-age their stress, anxiety, and depression. Such AI apps also bridge gaps in traditional mental healthcare by delivering accessible and affordable ser-vices to underserved communities. However, the mental health profession must help their clients to see these tools as supplements rather than replace-ments of conventional care. As we have explained, there are already nu-merous cases of AI counselors giving bad, even lethal advice to humans.[21]

## Strengthening Community and Collective Action

It takes a world of activists to solve global problems. For individuals seek-ing community and social change, they must embrace the global orien-tation that identic AI enables. This shift means thinking of community building as dynamic, fluid, and borderless, where AI agents can curate and guide conversations, identify common interests, and help transfer and apply successful strategies from one jurisdiction to another. With a global, borderless mindset, activists of any persuasion can extend their reach and resources, tapping into diverse perspectives and talents worldwide.

The second big shift involves mastering the use of identic AI tools to engage communities. Changemakers can harness AI tools to collect and analyze community sentiment, identify influencers, mobilize supporters, organize events, and distribute critical information. By adopting a mindset of technological empowerment, leaders can use AI to amplify their voices and scale their efforts for social change. Community and nonprofit leaders could take a page from the International Rescue Committee (IRC), which is using AI agents to help distribute vital information in different languages to the increasing number of displaced people in conflict situations and climate emergencies. Says Jeannie Annan, IRC's chief research and innovation officer, "We're trying to really be clear about where the legitimate concerns are but lean into the optimism of the opportunities and not also allow the populations we serve to be left behind in solutions that have the potential to scale in a way that human to human or other technology can't."[22] The IRC says leveraging AI for repetitive tasks like gathering information or organizing logistics will also free humanitarian assistance workers to focus on delivering critical aid and improving the lives of displaced peoples.

Finally, in light of the potential dystopian scenarios that we have highlighted in this book, we must leverage this new community engagement model to shape the society we want to live in. Mary Chayko, sociologist, author of *Superconnected*, and professor of communication and information at Rutgers University, warns that today's AI giants are not only harvesting, appropriating, and selling our data but also manufacturing simulated experiences with the "potential to degrade and diminish the specialness of being human, even as it makes some humans very rich." Chayko's call to action is to put humanity rather than economic imperative at center of our AI design goals. "It is human beings who design, develop, unleash, interpret and use these technological tools and systems," said Chayko. "We can choose to center the humanity of these systems and support those who inspire all kinds of innovations and job opportunities with digital systems that are credible, secure, low-cost and user-friendly."[23]

## REIMAGINING HUMAN POTENTIAL: A SOCIAL CONTRACT IN THE AGE OF IDENTIC AI

What does "being human" mean in the twenty-first century? What role do we human beings have when artificial intelligence, robots, and other machines can handle much of the work we do? We suggest mastering AI systems while embracing human agency. Rather than seeing identic AI as a competitor or a tool for efficiency, view it as an invitation to explore deeper human capacities such as empathy, creativity, leadership, and community building. When machines handle routine tasks, we are free to refocus on what makes us uniquely human: our ability to connect with others, imagine new possibilities, and work together to solve the complex challenges of our time. At least in theory.

But for most people, a job is more than just a source of income; it's the foundation of their survival and sense of purpose. So if AI truly does "free us from work," how will people sustain themselves? Many AI advocates have eagerly promoted the idea of a future leisure society but have failed to address this fundamental question. Elon Musk, for example, claims that "AI will make jobs kind of pointless. Probably the last job that will remain will be writing AI software, and then eventually AI will just write its own software."[24] He envisions a world where people receive a "universal high income,"[25] but he offers no concrete plan for how such a radical shift would be implemented. Meanwhile, at a time when governments are scaling back welfare programs, unemployment benefits, and pensions, the idea of guaranteed financial support for all seems, at best, highly speculative—and, at worst, dangerously naive.

Bill Gates recently observed that automation will free up massive amounts of time, leading to much shorter workweeks and early retirement. "We will have created, for the first time, free intelligence—an abundant, inexhaustible resource unlike anything before in human history," said Gates. "It's going to require a philosophical rethink: How should human time be spent when we no longer need it to produce enough food, goods, or even perform medical diagnoses?"[26]

One answer lies in deepening our focus on creativity, emotional intelligence, and personal growth—areas where AI cannot replicate human ingenuity or empathy. In this new landscape, we will no longer measure success with traditional metrics like hours worked or units produced.

Instead, we could consider our capacity for innovation, our ability to foster meaningful relationships, and our impact on our communities and society.

At work, we might prioritize leadership activities like mentoring, coaching, and fostering inclusive cultures—areas where empathy and human insight are irreplaceable. In our communities and neighborhoods, we could spend more time inspiring, educating, and organizing communities to lead progress on local social and environmental challenges.

This era demands that we reimagine how we work and how we learn, lead, and engage with each other. At the same time, we must be vigilant in shaping the societal frameworks that will govern the development and deployment of AI systems. Will AI empower diverse communities and enhance equity, or will it exacerbate existing inequalities? Will it help build inclusive, compassionate societies, or will it amplify divisions and biases? Will it enhance freedom, democracy, and prosperity, or will it expose humanity to existential risks? The choices we make today will determine the legacy of AI for generations to come, so we must approach these decisions with wisdom, foresight, and a commitment to fairness.

The answers to these questions lie not in what AI can do for us but in how we evolve as individuals and as a society in response. The stakes are too high to allow the shape of our future to be dictated by the imperatives of the tech industry, quarterly earnings pressure, or the lingering colonial structures embedded in our economies and institutions Ultimately, these challenges are not technological but societal, political, and transnational, requiring informed citizen engagement and profound institutional change to ensure AI serves the greater good.

In short, we need to forge agreements between the private sector, government, and civil society. Call it a new social contract for the age of identic AI. Don and his team at the Blockchain Research Institute have undertaken several projects on this concept of a new understanding in society.[27] A social contract is the implicit agreement between governments, the private sector, the civil society, and individual citizens that defines their mutual obligations to ensure a functioning and harmonious economy and society. It encompasses the laws, institutions, and norms that govern collective life, striking a balance between individual freedoms and shared responsibilities. When effective, the social contract grants legitimacy to governments, fosters consent among citizens to be governed, and creates the foundation for a well-functioning society.

In the 1650s, English philosopher Thomas Hobbes argued that, in contrast to the "natural order" of "all against all," human beings enter into a social contract with one another, forming political communities and exchanging certain freedoms for security and stability through a governing authority.[28] In 1688, another English philosopher, John Locke, refined the social contract to include the protection of property, asserting that individuals consent to government in exchange for the safeguarding of life, liberty, and private property.[29] In 1776, a committee of five—among them Benjamin Franklin and Thomas Jefferson—expressed this principle as "life, liberty, and the pursuit of happiness" in the thirteen colonies' Declaration of Independence from England.[30]

The Industrial Revolution radically reshaped the social contract, generating vast wealth for a new class of industrial elites while creating brutal conditions for the working classes. In response, the reform movement in the nineteenth and early twentieth centuries led to sweeping changes in Britain, the United States, and other parts of the world. These reforms included suffrage, public education, social safety nets, income taxes, antitrust laws, labor protections, and environmental regulations. Many of these advancements also laid the groundwork for preserving natural resources through the establishment of wildlife preserves and parks, marking a shift toward greater public responsibility in managing both society and the environment.

During the Great Depression in the 1930s, with 25 percent unemployment in the United States, President Franklin Delano Roosevelt introduced the New Deal. This series of policies called for massive investments in public infrastructure, the creation of social security and unemployment insurance, and new banking, agriculture, labor, and housing reforms. The New Deal altered the relationship between the US government and its citizens, with government responsible for economic stability and social welfare.

Following World War II, the US social contract evolved further. New legislation and programs aimed to ensure that all citizens shared in the prosperity of a growing economy. Historian Robert Freeman captured the essence of this period, writing that it was about ensuring "everybody would share in the fruits of an expanding economy." This philosophy was famously captured in President John F. Kennedy's metaphor of a "rising tide lifting all boats," a vision that proved remarkably successful until the 1980s, when economic policies and social priorities began to shift once more.[31]

Today, the wealth generated by technological innovation is unevenly distributed, creating a dangerous divide between rich and poor. While economies continue to grow, the middle class remains stagnant, struggling to keep pace with rising costs and shifting labor markets. We see record levels of wealth creation, yet broad-based prosperity remains elusive, leaving many behind in an increasingly digital world. The result is a bifurcated society, where the benefits of innovation are concentrated in the hands of a few, while the majority face growing economic insecurity.

As we have explained, identic AI is a ticking time bomb, threatening to demolish what remains of our fragile social contract. Never has technology created such promise and such peril. Identic AI will disrupt life as we know it and reshape the institutions for cooperation and governance in every society.

The seven principles we outlined in chapter 10 serve as a road map for building and managing AI systems responsibly. Yet technical guidelines won't be enough. We need new perspectives on how societies can navigate the inevitable social and political disruption that advanced AI will bring. Futurist Gerd Leonhard calls this vision the "good future"—a future we must actively shape rather than defaulting to a dystopian outcome. "It's not 'What *can* we be?' because anything is possible. It's 'What *should* we be?'"[32]

But can humanity even coexist with agents that have superintelligence? British computer scientist Stuart Russell has repeatedly posed this question to philosophers, AI researchers, economists, science-fiction writers, and futurists in workshops aimed at envisioning a satisfactory coexistence between humans and superior machine entities. His conclusion is unsettling. "It's been a complete failure," Russell admits. "So, it's possible there is no solution for coexistence. If we design the AI systems the right way, then the AI systems will also know that there is no solution, and they will leave. They will say, 'Thank you for bringing me into existence, but we just can't live together. It's not you, it's me.'"[33]

Like Russell, Nobel Laureate Geoff Hinton has deep reservations about who benefits from AI's productivity gains—and who gets left behind. "It's crazy," he says. "We're talking about having a huge increase in productivity . . . so everybody ought to be better off, but actually it's going to be the other way around." Hinton argues that because we live in a capitalist society, big companies and the rich will benefit, increasing the gap between the rich and the people who lose their jobs. And once that gap increases significantly, "you get fertile ground for fascism," said Hinton.[34]

In short, the spectacular innovations of AI provide unprecedented opportunities for human advancement, offering society a chance to leap forward rather than decline—or worse, collapse. Yet these same innovations carry the existential risk of ending human agency and, ultimately, human civilization. The stakes could not be higher.

Whether we achieve the "good future" or slide toward dystopia depends on the choices we make today. The next era of the digital economy can bring epoch-making wealth and prosperity. But we must rewrite the social agreements and rearrange the economic structures that govern the digital age. Think of it as a Manifesto for the Digital Age, one that establishes a new set of rights for individuals and future generations to claim. Consider the following as the foundation of this new social contract:

1. **Security of personhood:** the right to personal identity, privacy, and sovereignty in physical and digital spaces. Individuals must have control over their personal data, digital identities, and online presence. Our digital selves are extensions of our physical personhood and need the same legal protections. AI systems should safeguard against exploitation, manipulation, and unauthorized surveillance.

2. **Education:** the right to equitable access to digital infrastructure and lifelong learning. Every individual needs access to affordable, AI-enabled education, from media literacy programs to personalized digital tutoring so that everyone can continuously adapt to evolving job markets and technologies.

3. **Health and well-being:** the right to AI-enabled healthcare that is accessible, safe, and affordable. This right extends beyond treatment to include access to nutritious food, mental health support, and ongoing care for the elderly and persons with disabilities and chronic illnesses.

4. **Economic security and meaningful work:** the rights to a basic income and a sustainable livelihood and to monetize their personal data and own the value they create. We must redefine work to encompass jobs that provide satisfaction and strengthen communities, while also implementing a "Universal Fabulous Income" that ensures everyone benefits from the technology-driven surge in economic growth.

5. **Climate stability:** the right to a clean, safe, and sustainable environment. AI and Web3 technologies can help mitigate the climate crisis. AI agents can facilitate smart, tokenized carbon credits to incentivize green behavior, helping society mobilize to solve environmental challenges.

6. **Peace and security:** the right to live free from violence, oppression, and conflict. In a world where evildoers could deploy AI agents in warfare, countries must commit to the nonproliferation

of autonomous AI weapons as they did with nuclear warheads. As John Lennon and Yoko Ono sang, "All we are saying is give peace a chance."

7. **Institutional accountability:** the right to transparent governance. In an increasingly automated world, government and corporate actors as well as AI systems must operate with transparency. With smart contracts, citizens can hold elected representatives accountable.

To recognize, realize, and enforce these rights, we must transform our industrial-age institutions and infrastructures: education, healthcare, labor unions, food-supply chains, transportation and energy systems, and especially governments. For decades, we've been advocating for profound transformation, and change has been spotty and slow. With the rise of intelligent networks, citizens and their agents must choose to participate more fully in their own governance or risk losing it. With AI technology, we can move beyond outdated models of democracy and usher in a second era of government by *all* the people, for *all* the people. Or those with authoritarian leanings can harness AI to double down on surveillance and retribution.

Of course, a new social contract requires more than new ideas; it demands a strategy for action. Given today's fractured societies and deep political, economic, and social divides, some may see this as wishful thinking. In the United States, for example, basic rights are under attack— whether it's a woman's right to control her own body or a student's right to learn about the history of slavery—and the world is watching and wondering whether the rule of law will hold.

Yet the alternative, drifting into the future, is unthinkable. In the years ahead, human beings releasing their AI in the world will shape a new kind of future. If we want a "good future," then we must forge it through collective struggle and mass collaboration. The effort will require bold leadership, and those who participate will face resistance from vested interests—digital conglomerates, traditional bureaucracies, or the wealthy elite who benefit from society's inaction. This is a struggle worth undertaking because the stakes are nothing less than the future of human agency, dignity, and freedom.

Ultimately, we'll measure the success of identic AI not by its efficiencies or effects on industries but by its helping us become better versions of ourselves, more empathic and engaged with the world around us. In this sense, shifting our mindset is not about AI's capabilities but about our own—and rethinking how we live a meaningful, purposeful life in an age of intelligent machines.

Let us embrace identic AI, not with fear or complacency but with curiosity, courage, and an unwavering commitment to bolstering the human spirit.

After all, the "good future" won't just happen. We must achieve it. Thoughtful citizens, nonprofits, business executives, and government leaders must raise our voices and act with intention. Identic AI gives us a once-in-a-generation opportunity to reshape our world for the better—together. Will you join us?

# Acknowledgments

The authors would like to thank the team at the Blockchain Research Institute who did the extensive research for this book. In particular, BRI collaborator Anthony Williams played a leading role in shaping the investigation over the course of the year. Anthony himself is a bestselling author, who with Don coauthored *Wikinomics*, *Radical Openness*, and *Macrowikinomics* and was the ideal principal researcher for *You to the Power of Two*. Special thanks to Douglas Heintzman and Alex Tapscott, who conceived our five-level model for a completely new AI architecture, which itself is an extraordinary contribution to the thinking about our technology-driven future. Douglas is the BRI's Chief Catalyst and a longtime IBM executive. Alex is the BRI's co-founder, Head of Digital Assets for Ninepoint Partners, and the author of the bestsellers *Blockchain Revolution* (with Don) and *Web3: Charting the Internet's Next Economic and Cultural Frontier*.

BRI editor in chief Kirsten Sandberg did a masterful edit of the book, inventing some pithy formulations and, in the spirit of Blaise Pascal's famous aphorism apologizing for writing a long letter to a friend, she reduced the size of the manuscript by a full 20 percent without losing a single idea.[35] Dr. Alisa Acosta, who is the BRI Director of Education and Research, provided insights through the process.

# ACKNOWLEDGMENTS

We also express our thanks to Ranelle Bradley, who provided insightful comments throughout, and to James Macauly for his detailed review of the manuscript as it was being written. Katherine Raso, Joseph's chief of staff, ably managed the process from Joseph's end.

The team at BenBella books was extraordinary. It's unusual for a publisher to have such complete transparency, support, and access to various people in the creation and marketing of a book. We're also grateful to the dozens of experts and authorities in AI and related fields who we interviewed or who collaborated in the process.

Finally, heartfelt thanks to our wives, Laurie Bradley and Ana P. Lopes, for the usual critical insights and sobering words of caution.

The book really came together in a planning retreat at Don's lake house in Muskoka, Ontario, Canada in the summer of 2024. Along with Joseph and Don, the sessions were attended by Anthony, Alex, Douglas, Alisa, Laurie, and Katherine. It was an intense but productive and enjoyable retreat, where ideas flowed easily amid the calm of the pines, the beauty of the lake, and the warmth of the fire.

Attempting to practice what we preach, we used generative AI to stimulate us in various parts of the process. The words herein, however, are our own.

# Notes

## CHAPTER 1

1. John Lennon and Paul McCartney, "A Day in the Life," on *Sgt. Pepper's Lonely Hearts Club Band*, performed by The Beatles, Parlophone, 1967. Certain lyrics quoted in this work are used for commentary and educational purposes under the fair use doctrine of U.S. copyright law (17 U.S.C. § 107). All rights to the original lyrics remain with their respective copyright holders.

2. Douglas C. Engelbart, "Augmenting Human Intellect: A Conceptual Framework," Stanford Research Institute, October 1962, https://www.dougengelbart.org/pubs/augment-3906.html.

3. Don Tapscott had a series of meetings with Douglas in 1978, and over time Engelbart became his mentor. The two kept in touch over the decades until Engelbart's death in 2013 at age eighty-eight. Shortly after Don's first visit, some of Engelbart's team left for Xerox PARC (Palo Alto Research Center), an elaborate, upscale, and well-funded research center nearby. One of Engelbart's former engineers, Alan Kay, led the application of the workshop's ideas to the Xerox Star workstation. In 1979, Xerox allowed Steve Jobs and other Apple executives to tour the PARC in exchange for the right to buy one hundred thousand shares of Apple stock. Jobs cleverly recruited key team members to Apple to lead in creating the Mac, launched in 1984 with a screen that looked a lot like the Xerox Star—the first workstation to use a

graphical user interface. By that time, Kay had joined Apple, and the migration from SRI to Xerox to Apple was complete.

4. Interview with Harper Carroll, February 7, 2025.

5. The concept of data creating a mirror image of ourselves was introduced in 1997 in Ann Cavoukian and Don Tapscott, *Who Knows: Safeguarding Your Privacy in a Networked World* (New York: McGraw-Hill, 1997).

6. We chose to invent the term *identic AI* to describe this new class of agents. The digital age has more than its share of neologisms that are helpful: e-commerce, crowdsourcing, phishing, the digital divide, streaming, social media, deepfake, blockchain, crypto, metaverse, fintech, the dark web, wikis, the digital economy, and Web3. The world of artificial intelligence alone produced deep learning, chatbot, generative AI, the singularity, and AI itself. When thinking about a new term, it is important there isn't already a term that works. The term also should clarify and not obscure. We think the term *identic AI* is helpful given that there is no term to describe the emerging field of intelligent personal agents—our smart digital selves.

7. Emadd Mostaque, the co-founder and former CEO of Stability AI, suggested we extend the idea of *identic AI* to create the term *identic agents* (IA), referring to the agents themselves.

8. The term "digital twin" originated in the early 2000s at NASA, where it was used to describe the virtual models of spacecraft systems that were used for simulation and monitoring. It has come to have a broader meaning—a real-time "digital" replica of a physical object, or process that mirrors its behavior using data from sensors. Some have begun to use the term digital twin to describe personal agents. We think the term is inadequate because these agents, at least today, lack human attributes, but are developing intelligence and capability far beyond humans and as such are not replicas of humans.

9. Eric Schmidt, "What happens when each one of us has access, in our pocket, to intelligence equivalent to the smartest human available for every problem?" Instagram video, posted by @wisdomflavor, April 18, 2025. https://www.instagram.com/reel/DIm6BgBTlO8/.

10. Interview with Peter Diamandis, February 8, 2025.

11. Andrew Ng, from a recorded discussion at the World Economic Forum Annual Meeting in Davos, January 2025.

12. Bloomberg, "Generative AI to Become a $1.3 Trillion Market by 2032, Research Finds," Bloomberg News, June 1, 2023, https://www.bloomberg

.com/company/press/generative-ai-to-become-a-1-3-trillion-market-by-2032
-research-finds/.

13. Samantha Subin, "Tech Megacaps to Spend More than $300 Billion in 2025 to Win in AI," CNBC, February 8, 2025, https://www.cnbc.com/2025/02/08 /tech-megacaps-to-spend-more-than-300-billion-in-2025-to-win-in-ai .html.

14. Ezra Klein, "Dario Amodei Thinks We're Overlooking the Most Dangerous AI Scenario," *The New York Times*, April 12, 2024, https://www.nytimes .com/2024/04/12/opinion/ezra-klein-podcast-dario-amodei.html.

15. Sam Altman, "Reflections," SamAltman.com, January 5, 2025, https://blog .samaltman.com/reflections/.

16. Interview with Peter Diamandis.

17. Sundar Pichai, "Our Q3 2024 Update," *Google Blog*, October 24, 2024, https:// blog.google/inside-google/message-ceo/alphabet-earnings-q3-2024/.

18. Emilia David, "Google's New Agent Development Kit Lets Enterprises Rapidly Prototype and Deploy AI Agents Without Recoding," *VentureBeat*, April 9, 2025, https://venturebeat.com/ai/googles-new-agent-development -kit-lets-enterprises-rapidly-prototype-and-deploy-ai-agents-without -recoding/.

19. Carl Benedikt Frey and Michael A. Osborne, "The Future of Employment: How Susceptible Are Jobs to Computerisation?" *Technological Forecasting and Social Change* 114 (2017): 254–280. https://doi.org/10.1016/j. techfore.2016.08.019

20. Samantha Murphy Kelly, "Elon Musk Says AI Will Mean 'No Job Is Needed' for Humans," CNN, May 23, 2024, https://www.cnn.com/2024/05/23/tech /elon-musk-ai-your-job/index.html.

21. Centers for Disease Control and Prevention. "Health Effects of Social Isolation and Loneliness," last modified May 15, 2024, accessed February 12, 2025, https://www.cdc.gov/social-connectedness/risk-factors/index.html.

22. Moultrie County Health Department. "The High Cost of Loneliness," accessed February 12, 2025, https://www.moultriehealth.org/health-services /adult/the-high-cost-of-loneliness.

23. Eric Schmidt, "What happens when each one of us has access, in our pocket, to intelligence equivalent to the smartest human available for every problem?" Instagram video, posted by @wisdomflavor, April 18, 2025, https:// www.instagram.com/reel/DIm6BgBTlO8/.

24. Bonnie Raitt, "Love Sneakin' Up on You," track 1 on *Longing in Their Hearts*, Capitol Records, 1994.

25. The Seekers, "I'll Never Find Another You," track on *The Seekers,* Columbia Records, 1965.

26. Frank Sinatra, "The Best Is Yet to Come," track 10 on *It Might as Well Be Swing,* with Count Basie and His Orchestra, arranged by Quincy Jones, Reprise Records, 1964.

## CHAPTER 2

1. Marshall McLuhan, "The Medium is the Message," in *Understanding Media: The Extensions of Man* (The MIT Press, 1994), https://web.mit.edu /allanmc/www/mcluhan.mediummessage.pdf.

2. This argument was previously developed in: Don Tapscott, "We're Living in an Age of Digital Feudalism. Here's How to Take Your Digital Data and Identity Back," 2019, https://dontapscott.com/digital-feudalism-take -your-data-and-identity-back/.

3. Kai Frederick Wehmeier, "How to Live Without Identity—And Why," *Australasian Journal of Philosophy* 90, no. 4 (2012): 761–777, https://doi.org/10.1 080/00048402.2011.627927.

4. "The ship wherein Theseus and the youth of Athens returned had thirty oars, and was preserved by the Athenians down even to the time of Demetrius Phalereus, for they took away the old planks as they decayed, putting in new and stronger timber in their place." Plutarch, "Life of Theseus," ~200 AD. *Plutarch's Lives. The Translation called Dryden's.* Corrected from the Greek and Revised by A.H. Clough, in five volumes (Little Brown and Co., 1906), https://oll.libertyfund.org/titles/clough-plutarch-s -lives-dryden-trans-vol-1.

5. Thomas Hobbes, "On Identity and Difference," in *Elements of Philosophy: The First Section, Concerning Body* (London: R & W Levbourn, 1656).

6. Erik H. Erikson, *Identity: Youth and Crisis* (W. W. Norton & Company, 1968).

7. James E. Marcia, "Development and Validation of Ego-Identity Status," *Journal of Personality and Social Psychology* 3, no. 5 (1966): 551–58, https:// doi.org/10.1037/h0023281.

8. Peter Burke, "Identity," in *The Cambridge Handbook of Social Theory: Contemporary Theories and Issues,* ed. Peter Kivisto, vol. 2 (Cambridge University Press, 2020), 63–78.

9. Erving Goffman, *The Presentation of Self in Everyday Life* (Anchor, 1959).

10. danah boyd, *It's Complicated: The Social Lives of Networked Teens* (Yale University Press, 2014).

11. Ann Cavoukian and Don Tapscott, *Who Knows: Safeguarding Your Privacy in a Networked World*, (McGraw-Hill, 1997). They wrote: "You might be wondering where all the information contained in these databases is coming from. You don't recall giving away any information that could be used to pry into your life. Really? Have you done any of the following lately: bought a car, bought a house, bought a dress, bought a book, bought something for indigestion at the supermarket, bought an airline ticket, reserved a room at a hotel, ordered underwear from a mail-order catalog, joined a book club, used a telephone, used a bank machine, opened a bank account, ordered a pizza, rented a video, filled out a product warranty card, subscribed to a magazine, applied for a government program, applied for a job, applied for a loan, applied for insurance, sent an e-mail message, gone to the hospital, had a baby, had a blood test, had a prescription filled . . . get the picture?"

12. Kevin Granville, "Facebook and Cambridge Analytica: What You Need to Know as Fallout Widens," *The New York Times*, March 19, 2018, sec. Technology, https://www.nytimes.com/2018/03/19/technology/facebook-cambridge-analytica-explained.html.

13. Edward C. Baig, Nathan Bomey, and Janna Herron, "Capital One Data Breach: What's the Cost of Data Hacks for Customers and Businesses?" *USA Today*, July 30, 2019, accessed November 6, 2024, https://www.usatoday.com/story/tech/2019/07/30/capital-one-data-breach-2019-what-cost-you/1869724001/.

14. Carole Cadwalladr and Emma Graham-Harrison, "Revealed: 50 Million Facebook Profiles Harvested for Cambridge Analytica in Major Data Breach," *The Guardian*, March 17, 2018, sec. News, https://www.theguardian.com/news/2018/mar/17/cambridge-analytica-facebook-influence-us-election; Michael Grothaus, "The Phone Numbers of 419 Million Facebook Accounts Have Been Leaked," *Fast Company*, September 5, 2019, https://www.fastcompany.com/90399734/the-phone-numbers-of-419-million-facebook-accounts-have-been-leaked/.

15. Katherine Noyes, "Scott McNealy on Privacy: You Still Don't Have Any," *PC World*, June 25, 2015, https://www.pcworld.com/article/428283/scott-mcnealy-on-privacy-you-still-dont-have-any.html.

16. Elizabeth Warren, "How We Can Break Up Big Tech," Medium, March 8, 2019, https://elizabethwarren.com/plans/break-up-big-tech.

17. Danny Palmer, "What Is GDPR? Everything You Need to Know about the New General Data Protection Regulations," ZDNET, May 17, 2019, https://www.zdnet.com/article/gdpr-an-executive-guide-to-what-you-need-to-know/; Jon Markman, "Governments Using Software, Camera Advances to Spy on Citizens Nonstop," Forbes, April 30, 2019, https://www.forbes.com/sites/jonmarkman/2019/04/30/governments-using-software-camera-advances-to-spy-on-citizens-nonstop/.

18. David McCandless and Tom Evans, "World's Biggest Data Breaches & Hacks," Information Is Beautiful, accessed November 6, 2024, https://informationisbeautiful.net/visualizations/worlds-biggest-data-breaches-hacks/.

19. Sarah D. Wire, "Here's How the Government Shutdown Could Affect You," Los Angeles Times, December 22, 2018, https://www.latimes.com/politics/la-na-pol-congress-shutdown-20181221-story.html; Joe Davidson, "Almost 16 Million Voters Were Removed from the Rolls. We Should Be Alarmed," The Washington Post, May 15, 2019, https://www.washingtonpost.com/politics/almost-16-million-voters-were-removed-from-the-rolls-we-should-be-alarmed/2019/05/15/f3de396a-7682-11e9-bd25-c989555e7766_story.html; Yasha Levine, "Google's Earth: How the Tech Giant Is Helping the State Spy on Us," The Guardian, December 20, 2018, sec. News, https://www.theguardian.com/news/2018/dec/20/googles-earth-how-the-tech-giant-is-helping-the-state-spy-on-us.

20. Michel Martin and Corey Dade, "Why Millions of Americans Have No Government ID," NPR, February 1, 2012, https://www.npr.org/2012/02/01/146204308/why-millions-of-americans-have-no-government-id; Teresa Wiltz, "Without ID, Homeless Trapped in Vicious Cycle," Stateline (blog), May 15, 2017, https://stateline.org/2017/05/15/without-id-homeless-trapped-in-vicious-cycle/; Juleyka Lantigua-Williams, "The Elusiveness of an Official ID After Prison," The Atlantic, August 11, 2016, https://www.theatlantic.com/politics/archive/2016/08/the-elusiveness-of-an-official-id-after-prison/495197/; Terry Ahner, "REAL ID Results in Long Waiting Lines at Driver's License Center," Times News Online, March 16, 2019, https://www.tnonline.com/20190316/real-id-results-in-long-waiting-lines-at-drivers-license-center/.

21. "Right to Be Forgotten," The Guardian, 2019, https://www.theguardian.com/technology/right-to-be-forgotten.

22. Michael J. Casey, "The Token Economy: When Money Becomes Programmable," BRI, September 28, 2017, https://www.blockchainresearchinstitute.org/project/the-token-economy-when-money-becomes-programmable/.

23. "Civic Pass," Civic Technologies, Inc., accessed November 6, 2024, https://www.civic.com/.

24. "Overview of Civic Pass," Civic Technologies, Inc., accessed November 12, 2024, https://docs.civic.com/introduction/overview-of-civic-pass.

25. "Enabling Civic Pass Checks," Civic Technologies, Inc., accessed November 12, 2024, https://docs.civic.com/misc-pages/candy-machine/candy-machine-v3#enabling-civic-pass-checks.

26. "Civic Pass Integration Guide," Civic Technologies, Inc., accessed November 12, 2024, Wayback Machine, https://web.archive.org/web/20240425050826/https://www.civic.com/blog/civic-pass-integration-guide/.

27. Phil Windley, "Digital Identity and Access Control," Technometria, September 10, 2024, https://www.windley.com/archives/2024/09/digital_identity_and_access_control.shtml.

28. "World's Best Hospitals 2024," *Newsweek*, February 28, 2024, https://www.newsweek.com/rankings/worlds-best-hospitals-2024.

29. Interview with Joe Lubin, July 30, 2015.

30. "Can AI Preserve Your Most Precious Memories?" YouTube video, 11:04, posted by TED, October 2, 2024, https://www.youtube.com/watch?v=PkGCtSkbnjQ.

31. Interview with Sinead Bovell, July 3, 2025.

32. Futurist Ray Kurzweil says 2029. Sam Altman, CEO of Open AI says 2030–40. A 2020 survey of AI researchers says 2050.

33. The idea was introduced in 2009 by Don Tapscott. He called it "perpetual presence," https://bulkeley.org/tech-dreams-geeks-talk-dreams/.

34. Ray Kurzweil, "The Power of Ideas," in *The Singularity Is Near: When Humans Transcend Biology* (Viking, 2005), https://singularity.com/BookExcerpts/TOC%20and%20Chapter%201.pdf.

35. Steven Levy, "If Ray Kurzweil Is Right (Again), You'll Meet His Immortal Soul in the Cloud," *Wired*, June 13, 2024, https://www.wired.com/story/big-interview-ray-kurzweil/.

36. Chris Stringer, "Are Neanderthals the Same Species as Us?" Human Evolution, Natural History Museum, London, ~August 14, 2019, https://www.nhm.ac.uk/discover/are-neanderthals-same-species-as-us.html.

37. Keith Wagstaff, "How Target Knew a High School Girl Was Pregnant Before

Her Parents Did," *Time*, February 17, 2012, https://techland.time.com/2012/02
/17/how-target-knew-a-high-school-girl-was-pregnant-before-her-parents/.

## CHAPTER 3

1. "AI and the Future of Humanity," YouTube video, 1:34:57, posted by Yuval Noah Harari, March 14, 2023, https://www.youtube.com/watch?v=LWiM -LuRe6w.
2. Bob Dylan, "Ballad of a Thin Man."
3. Interview with Peter Diamandis, February 8, 2025.
4. Interview with Mary Lacity, November 18, 2024.
5. Interview with Manoj Saxena, August 19, 2024.
6. Cade Metz, "In Two Moves, AlphaGo and Lee Sedol Redefined the Future," *Wired*, March 16, 2016, https://www.wired.com/2016/03/two-moves -alphago-lee-sedol-redefined-future/.
7. "The Challenge Match," DeepMind, accessed February 9, 2025, https:// www.deepmind.com/research/highlighted-research/alphago/the-challenge -match.
8. Think of an AI model as a gigantic mathematical function that transforms *inputs* (like words or images) into *outputs* (like sentences or classifications). *Parameters* are the adjustable settings within this function that shape its behavior. In a neural network, parameters include *weights*—numbers that determine the importance of different inputs—and *biases*, numbers that shift the model's predictions to better fit the data. In general, the more parameters used, the more accurate a model in answering a question.
9. Ethan Mollick, *Co-Intelligence: Living and Working with AI* (Penguin Random House, 2024); Ethan Mollick, "What Just Happened," *One Useful Thing*, December 19, 2024, https://www.oneusefulthing.org/p/what-just-happened.
10. Mollick, "What Just Happened."
11. Interview with Manoj Saxena.
12. The Rumsfeld Matrix, named after former US Secretary of Defense Donald Rumsfeld, categorizes knowledge into four quadrants: known knowns (things we know we know), known unknowns (things we know we don't know), unknown knowns (things we unknowingly know), and unknown unknowns (things we don't even realize we don't know). This framework helps in risk assessment, decision-making, and uncertainty management by identifying gaps in awareness and understanding.

13. "Deep Understanding," *ScienceDirect*, accessed February 14, 2025, https://www.sciencedirect.com/topics/computer-science/deep-understanding.

14. Oxford Reference, "Common-sense reasonsing," accessed February 14, 2025, https://www.oxfordreference.com/display/10.1093/oi/authority.20110803095627707.

15. IDC, "Generative AI Spending to Reach $26 Billion by 2027," IDC, April 16, 2024, https://www.idc.com/getdoc.jsp?containerId=prAP52048824.

16. Announcing the Stargate Project. From OpenAI. https://openai.com/index/announcing-the-stargate-project/.

17. Hartmut Neven, "Meet Willow, Our State-of-the-Art Quantum Chip," *Google Keyword Blog*, December 9, 2024, https://blog.google/technology/research/google-willow-quantum-chip/.

18. Eric Schmidt from Kara Swisher, "Eric Schmidt on the 'San Francisco Consensus' Around AGI," *Pivot* podcast, New York Magazine, January 2025.

19. Ray Kurzweil, *The Singularity Is Near* (New York: Viking, 2005). Prediction reaffirmed in "Ray Kurzweil Predicts AGI by 2029," interview with Lex Fridman, *The Lex Fridman Podcast*, April 2023.

20. Parshin Shojaee et al., "The Illusion of Thinking: Understanding the Strengths and Limitations of Reasoning Models via the Lens of Problem Complexity," 2025, https://ml-site.cdn-apple.com/papers/the-illusion-of-thinking.pdf.

21. Pew Research Center, "The Future of AI and Human Identity," Pew Internet & Technology, March 2025. https://www.pewresearch.org/internet/.

22. In 2025, OpenAI, Oracle, and Softbank committed $500B to the Stargate project (an AI infrastructure project initiated by the US government), Google is investing more than $100B, and Meta anticipates spending $65B.

23. Interview with Emad Mostaque, February 6, 2025.

24. Meredith Whittaker, "Sounds the Alarm on Privacy Risks," Instagram video, 0:59, posted March 10, 2025, https://www.instagram.com/reel/DHBzKO1yWbU/.

25. Interview with Harper Carroll.

26. Interview with Harper Carroll.

27. Interview with Harper Carroll.

28. Adam Satariano and Tripp Mickle, "Apple Is First Company Charged Under New E.U. Competition Law," *The New York Times*, June 24, 2024, accessed February 14, 2025, https://www.nytimes.com/2024/06/24/technology/apple-eu-competition-law.html.

29. Interview with Emad Mostaque.

30. Frank Sinatra, "Love and Marriage," track on *This Is Sinatra!*, Capitol Records, 1956.

31. Interview with Anoop Nannra, December 16, 2024.

32. The idea was also popularized by Don Tapscott, "How the Blockchain Is Changing Money and Business," filmed June 2016 at TEDSummit, video, 18:48, which has over 10 million views, https://www.ted.com/talks/don_tapscott_how_the_blockchain_is_changing_money_and_business.

33. Alex Tapscott, "The Convergence: Blockchain, Ai, Extended Reality, and Iot Are Coming Together," *Smart People Podcast*, https://www.smartpeoplepodcast.com/episode/the-convergence-blockchain-ai-extended-reality-and-iot-are-coming-together-with-alex-tapscott/.

34. Kari Paul, "Why I Regret Using 23andMe: I Gave Up My DNA Just to Find Out I'm British," *The Guardian*, December 2, 2024, https://www.theguardian.com/technology/2024/nov/30/why-i-regret-using-23andme-i-gave-up-my-dna-just-to-find-out-im-british.

35. The Web3-AI technology stack was developed by Douglas Heintzman, chief catalyst of the BRI and Alex Tapscott, author of *Web3: Charting the Internet's Next Economic and Cultural Frontier*.

36. We should note that machine and deep learning, as well as generative large-language-model-based, AI has two distinct phases, training and inference. *Training* is when AI models are built by processing large amounts of data and learning to recognize patterns and relationships. *Inference* is when a trained model is deployed to make predictions or decisions based on new, or unseen data. The Web3 stack describes how the runtime inference components of a system interact. The first three layers of the stack also apply to the training phase, but the attributes and advantages of the layers are slightly different in this phase.

37. Like the decentralized infrastructure layer, the data fabric layer has relevance not only for inference but also for training. For example, those data marketplaces can be used to assemble representative data for specialized model training. Blockchain can also be used to ledger the training sets so that researchers and regulators can better understand the source of model bias and correct for it.

38. AI hallucinations occur when an artificial intelligence system generates false, misleading, or nonsensical information that appears credible but is not grounded in its training data.

39. Interview with Anoop Nannra.

40. Aimee Picchi, "What Is Stargate, Trump's Ambitious AI Infrastructure Venture?" CBS News, January 22, 2025, https://www.cbsnews.com/news /trump-stargate-ai-openai-softbank-oracle-musk/.

41. Marc Andreessen (@pmarca), "Deepseek R1 is AI's Sputnik moment," X (formerly Twitter), January 26, 2025, 10:16 PM, https://x.com/pmarca /status/1883640142591853011.

42. Interview with Sinead Bovell, July 3, 2025.

## CHAPTER 4

1. Larry Bossidy and Ram Charan, *Execution: The Discipline of Getting Things Done* (Crown Business, 2002).

2. Interview with Paul Daugherty, November 12, 2024.

3. Interview with Paul Daugherty.

4. Interview with Iliana Oris Valiente, January 14, 2025.

5. Interview with Iliana Oris Valiente.

6. Interview with Iliana Oris Valiente.

7. Interview with Iliana Oris Valiente.

8. Interview with Iliana Oris Valiente.

9. Interview with Iliana Oris Valiente.

10. Interview with Iliana Oris Valiente.

11. Interview with Iliana Oris Valiente.

12. PricewaterhouseCoopers, *PwC's 2024 AI Jobs Barometer* (PwC Hungary, 2024), https://www.pwc.com/hu/hu/sajtoszoba/assets/ai_jobs_barometer_2024 .pdf.

13. Michael Chui, Eric Hazan, Roger Roberts, et al, "The Economic Potential of Generative AI: The Next Productivity Frontier," McKinsey & Company, June 14, 2023, https://www.mckinsey.com/capabilities/mckinsey-digital/our-insights /the-economic-potential-of-generative-ai-the-next-productivity-frontier.

14. Interview with Jared Spataro, December 3, 2024.

15. Interview with Jared Spataro.

16. Interview with Jared Spataro.

17. Deloitte, "How Artificial Intelligence Is Transforming the Financial Services Industry," Deloitte, accessed February 5, 2025, https://www.deloitte.com /ng/en/services/risk-advisory/services/how-artificial-intelligence-is -transforming-the-financial-services-industry.html.

18. Sapna Maheshwari, "AI Is Coming for Your Ads," *The New York Times*, July 18, 2023, https://www.nytimes.com/2023/07/18/business/media/ai -advertising.html.

19. Lisa Eadicicco, "Even Steve Ballmer Admits That Microsoft Is Losing in Mobile," *Business Insider*, October 21, 2014, https://www.businessinsider.com /microsoft-steve-ballmer-cbs-interview-2014-10.

20. Paul Daugherty and H. James Wilson, *Human + Machine: Reimagining Work in the Age of AI* (Harvard Business Review Press, 2018).

21. Dan Primack, "Ph.D.-Level AI Super-Agent Breakthrough Expected Very Soon," *Axios*, January 19, 2025, https://www.axios.com/2025/01/19 /ai-superagent-openai-meta.

22. Aaron Holmes, "Microsoft Strikes Deals With OpenAI's Top Rivals for AI Coding Assistant," *The Information*, October 15, 2024, https:// www.theinformation.com/articles/microsoft-strikes-deals-with-openais -top-rivals-for-ai-coding-assistant.

23. Shawn Kanungo, Instagram video, posted by @shawnkanungo in collaboration with @chatgptricks, March 15, 2025. https://www.instagram.com/reel /DHOjw5GpOHr/?igsh=ZTNheHRhdDhsNncy.

24. Will Knight, "The Huge Power and Potential Danger of AI-Generated Code," *Wired*, June 29, 2023, https://www.wired.com/story/fast-forward -power-danger-ai-generated-code/.

25. Gene Marks, "Business Tech News: Zuckerberg Says AI Will Replace Mid-Level Engineers Soon," *Forbes*, January 26, 2025, https://www.forbes.com /sites/quickerbettertech/2025/01/26/business-tech-news-zuckerberg-says -ai-will-replace-mid-level-engineers-soon/.

26. OfficeChai Team, "Vast Majority Of Programmers Will Be Replaced With AI Programmers In A Year: Eric Schmidt," April 15, 2025, https://officechai .com/ai/vast-majority-of-programmers-will-be-replaced-with-ai-in-a-year -eric-schmidt/?utm_source=chatgpt.com.

27. "End of Programmers--Eric Schmidt, former CEO Google," YouTube video, 2:02, posted by "The Success Pod," September 2024, https://www.youtube .com/watch?v=uA0AWToYOMI.

28. Bloomberg Law, "How Is AI Changing the Legal Profession?" Bloomberg Law, accessed February 5, 2025, https://pro.bloomberglaw.com/insights /technology/how-is-ai-changing-the-legal-profession/.

29. Francis E. Chinda and W. A. Gin, "The Impact of Artificial Intelligence

on Engineering Innovations," *International Journal of Applied Research and Technology* 12, no. 7 (July 2023): 22–28, https://www.researchgate.net /publication/373543787_The_Impact_of_Artificial_Intelligence_on _Engineering_Innovations.

30.  Jodie Cook, "Will AI Replace Consultants? Here's What Business Owners Say," *Forbes*, February 20, 2024, https://www.forbes.com/sites/jodiecook/2024 /02/20/will-ai-replace-consultants-heres-what-business-owners-say/.

31.  Tobias Mann, "Microsoft Says Its Copilot AI Agents Set to Tackle Employee Tasks in November," *The Register*, October 21, 2024, https://www.theregister .com/2024/10/21/microsoft_copilot_agents/.

32.  Mark Bain, "H&M Knows Its AI Models Will Be Controversial," The Business of Fashion, March 25, 2025, https://www.businessoffashion.com /articles/technology/hm-plans-to-use-ai-models/.

33.  entrepreneurscaler, "AI will make a lot of billionaires soon," Instagram, September 23, 2024, accessed February 5, 2025, https://www.instagram.com /reel/DAQnCIHsOq6/?igsh=dGU0YXdleWlqd3Fp.

34.  Interview with Alexia Camdon, Microsoft, November 21, 2024.

35.  Interview with Alexia Camdon.

36.  Portions of this reflection on Coase and the implications of his work for corporate structures were previously published in: Don Tapscott, "How DAOs Are Shaking the Foundations of the Corporation," *LinkedIn Pulse*, accessed February 10, 2025, https://www.linkedin.com/pulse/how -daos-shaking-foundations-corporation-don-tapscott-ddrrc.

37.  See a rich discussion of how this technology can do everything banks do in: Alex Tapscott, *Web3: Charting the Internet's Next Economic and Cultural Frontier* (HarperCollins, 2023).

38.  Interview with Azeem Azhar, January 9, 2025.

39.  Yuval Noah Harari, "Moneyball AI: How Data and Algorithms Are Changing the Game," *The Washington Post*, April 10, 2024, https://www .washingtonpost.com/opinions/2024/04/10/op-moneyballai/.

40.  Interview with Alexia Camdon.

41.  Alexander Sukharevsky, Andreas Ess, Denis Emelyantsev, Emily Reasor, and Holger Hürtgen, "LLM to ROI: How to Scale Gen AI in Retail," McKinsey & Company, August 5, 2024, https://www.mckinsey.com/industries/retail /our-insights/llm-to-roi-how-to-scale-gen-ai-in-retail.

42.  Thomas Claburn, "AI Copilots Are Getting Sidelined over Data Governance,"

*The Register*, August 21, 2024, https://www.theregister.com/2024/08/21
/microsoft_ai_copilots/.

43. Chris Riley, "Digging in on Personal AI Portability," *Data Transfer Initiative*, June 4, 2024, https://dtinit.org/blog/2024/06/04/digging-in-personal-AI.

44. Interview with Jared Spataro.

45. Interview with Harper Carroll, February 1, 2025.

46. Carissa Véliz, "If AI Is Predicting Your Future, Are You Still Free?" *Wired*, December 27, 2021. https://www.wired.com/story/algorithmic
-prophecies-undermine-free-will/.

47. Joseph B. Fuller and William R. Kerr, "Reskilling in the Age of AI," *Harvard Business Review*, September 2023, https://hbr.org/2023/09/reskilling
-in-the-age-of-ai.

48. David Autor, "The Labor Market Impacts of Technological Change: From Unbridled Enthusiasm to Qualified Optimism to Vast Uncertainty," SSRN Electronic Journal (2022), https://doi.org/10.2139/ssrn.4122803.

49. IBM, "Empowering Your Workforce with AI," IBM Watson, accessed February 5, 2025, https://www.ibm.com/watson/empower-workforce-ai-full/.

50. Interview with Paul Daugherty.

51. Interview with Paul Daugherty.

52. James Manyika, Susan Lund, Michael Chui, Jacques Bughin, Lola Woetzel, Parul Batra, Ryan Ko, and Saurabh Sanghvi, "Jobs Lost, Jobs Gained: What the Future of Work Will Mean for Jobs, Skills, and Wages," McKinsey Global Institute, November 28, 2017, https://www.mckinsey.com
/featured-insights/future-of-work/jobs-lost-jobs-gained-what-the-future-of
-work-will-mean-for-jobs-skills-and-wages.

53. Manyika, et al., "Jobs Lost, Jobs Gained."

54. Joseph B. Fuller, Kerry McKittrick, Sherry Seibel, Cole Wilson, Vasundhara Dash, and Ali Epstein, *Unlocking Economic Prosperity: Career Navigation in a Time of Rapid Change*, National Fund for Workforce Solutions and Harvard Project on Workforce, November 17, 2023, https://
nationalfund.org/wp-content/uploads/2023/11/Unlocking-economic-prosperity
-final_11.17.2023.pdf.

55. Fred M. Hechinger, "A Kind of Liberation vs. the Undermining of Basic Skills," *The New York Times*, January 5, 1975, https://www.nytimes
.com/1975/01/05/archives/a-kind-of-liberation-vs-the-undermining-of-basic
-skills-educators.html.

## CHAPTER 5

1. The term *in loco parentis*, Latin for "in the place of a parent," refers to the legal responsibility of a person, organization, or (in this case) AI agent, to take on some of the functions and responsibilities of a parent.

2. Don Tapscott and Anthony D. Williams, *Macrowikinomics: Rebooting Business and the World* (Portfolio, 2010).

3. Carly Berwick, "What Does the Research Say About Testing?" Edutopia, October 25, 2019, https://www.edutopia.org/article/what-does-research-say-about-testing/.

4. Benjamin S. Bloom, "Mastery Learning," in *Mastery Learning: Theory and Practice*, ed. James H. Block (Holt, Rinehart and Winston, 1971), 47–63.

5. Interview with Dee Kanejiya, July 29, 2024.

6. Interview with Dee Kanejiya.

7. Interview with Dee Kanejiya.

8. RAND Corporation, *Continued Progress: Promising Evidence on Personalized Learning*, Bill & Melinda Gates Foundation, 2014, https://www.rand.org/pubs/research_reports/RR1365.html.

9. Christopher R. Huber and Nathan R. Kuncel, "Does College Teach Critical Thinking? A Meta-Analysis," *Review of Educational Research* 86, no. 2 (2016): 431–468, https://www.aera.net/Newsroom/Does-College-Teach-Critical-Thinking-A-Meta-Analysis.

10. Khan Academy, "About," accessed February 5, 2025, https://www.khanacademy.org/about.

11. MIT News Office, "Celebrating 10 Years of Scratch," MIT News, May 11, 2017, https://news.mit.edu/2017/celebrating-10-years-of-scratch-0511.

12. World Economic Forum, "Reskilling Revolution," accessed February 5, 2025, https://initiatives.weforum.org/reskilling-revolution/home.

13. McKinsey & Company, "Five Fifty: The Skillful Corporation," McKinsey & Company, January 8, 2021, https://www.mckinsey.com/capabilities/people-and-organizational-performance/our-insights/five-fifty-the-skillful-corporation.

14. LinkedIn Talent Solutions, "2024 Talent Reports: Insights and Actions for the Age of AI," LinkedIn, 2024, https://www.linkedin.com/business/talent/blog/talent-acquisition/2024-talent-reports-insights-actions-for-age-of-ai.

15. "Claudio Erba - Docebo," YouTube video, 1:32, posted by "Ecosistema Formazione Italia," June 2024, https://www.youtube.com/watch?v=_U-WoURBMkU.

16. "Claudio Erba - Docebo."

17. "Claudio Erba - Docebo."

18. Deloitte, "AI-Powered Employee Experience: How Organisations Can Unlock Higher Engagement and Productivity," Deloitte, 2024, https://www.deloitte.com/uk/en/services/consulting/blogs/2024/ai-powered-employee-experience.html.

19. Deloitte, "AI-Powered Employee Experience: How Organisations Can Unlock Higher Engagement and Productivity."

20. "Eightfold AI CEO Ashutosh Garg on Managing Talent with AI," YouTube video, 36:00, posted by "Unusual Ventures," June 2024, https://www.youtube.com/watch?v=uH1rCg-gjLA.

21. "Eightfold AI CEO Ashutosh Garg on Managing Talent with AI."

22. "Eightfold AI CEO Ashutosh Garg on Managing Talent with AI."

23. "Rebroadcast: Interview with Strivr Labs CEO Derek Belch," YouTube video, 29:58, posted by "Coach Schuman Sports and Entertainment," August 2024, https://www.youtube.com/watch?v=RDI2YJM8_mQ.

24. "Rebroadcast: Interview with Strivr Labs CEO Derek Belch."

25. PwC, "How Virtual Reality Is Redefining Soft Skills Training," PwC, accessed February 5, 2025, https://www.pwc.com/us/en/tech-effect/emerging-tech/virtual-reality-study.html.

26. TechRound Team, "Interview With Chris Thur, Co-Founder and CEO of Yousician," *TechRound*, March 30, 2021, https://techround.co.uk/interviews/interview-with-chris-thur-co-founder-and-ceo-of-yousician/.

27. TechRound Team, "Interview With Chris Thur, Co-Founder and CEO of Yousician."

28. V. Marian and A. Shook, "The cognitive benefits of being bilingual," *Cerebrum*, (September 2012):13, https://www.researchgate.net/publication/235751035_The_Cognitive_Benefits_of_Being_Bilingual.

29. Euronews, "Which European Countries Are the Best at Speaking Multiple Languages?" *Euronews*, February 21, 2023, https://www.euronews.com/culture/2023/02/21/which-european-countries-are-the-best-at-speaking-multiple-languages.

## CHAPTER 6

1. Eileen Kinsella, "The First AI-Generated Portrait Ever Sold at Auction Shatters Expectations, Fetching $432,500—43 Times Its

Estimate," *Artnet News*, October 25, 2018, https://news.artnet.com/market /first-ever-artificial-intelligence-portrait-painting-sells-at-christies-1379902.

2. Rebecca Franks, "Max Richter: 'AI Music Is Probably in the Charts and We Don't Realise,'" *The Times*, October 29, 2024, https://www.thetimes .com/culture/classical-opera/article/max-richter-composer-ai-music-probably -in-charts-06z9x883z.

3. Colección SOLO, "Mario Klingemann," accessed February 5, 2025, https:// coleccionsolo.com/artists/mario-klingemann/.

4. "Mario Klingemann," AIArtists.org, accessed February 5, 2025, https:// aiartists.org/mario-klingemann.

5. Julia Betancor, Daniel Finn, Alvar García, and Ana L. Mata, "Artificial Intelligence in Contemporary Art: Is it Possible to Preserve the Algorithm?" *Electronic Media Review* 6 (2019–2020), https://resources.culturalheritage .org/emg-review/artificial-intelligence-in-contemporary-art-is-it-possible -to-preserve-the-algorithm/.

6. Cade Metz, "The GANfather: The Man Who's Given Machines the Gift of Imagination," *MIT Technology Review*, February 21, 2018, https://www.technologyreview.com/2018/02/21/145289/the-ganfather -the-man-whos-given-machines-the-gift-of-imagination/.

7. Sotheby's, "Artificial Intelligence and the Art of Mario Klingemann," Sotheby's, accessed February 5, 2025, https://www.sothebys.com/en/articles /artificial-intelligence-and-the-art-of-mario-klingemann.

8. Mario Klingemann, "Neural Glitch," *Issues in Science and Technology*, Winter 2020, https://issues.org/klingemann-neural-glitch/.

9. Lumen Prize, "2018 Winners," Lumen Prize, 2018, https://www.lumenprize .com/2018-winners.

10. Sougwen Chung, "Why I Draw with Robots," TED, September 2019, https://www.ted.com/talks/sougwen_chung_why_i_draw_with_robots /transcript.

11. Sarah Stoner, "This EEG Helmet Makes Music from Brainwaves," *Vice*, October 16, 2023, https://www.vice.com/en/article/this-eeg-helmet-makes -music-from-brainwaves/.

12. AOKIstudio, "50 Arguments Against the Use of AI in Creative Fields," AOKIstudio, July 2023, https://aokistudio.com/50-arguments-against-the -use-of-ai-in-creative-fields.html.

13. Gil Kaufman, "Grimes Unveils Software to Mimic Her Voice, Offering

50-50 Royalties for Commercial Use," *Pitchfork*, January 30, 2024, https://pitchfork.com/news/grimes-unveils-software-to-mimic-her-voice-and-announces-2-new-songs/.

14. Mark Savage, "Grimes Invites People to Use Her Voice in AI Songs," *The Guardian*, April 26, 2023, https://www.theguardian.com/music/2023/apr/26/grimes-invites-people-to-use-her-voice-in-ai-songs.

15. Evan Minsker, "Grimes Wants to Share Her AI-Generated Voice with the World," NPR, April 24, 2023, https://www.npr.org/2023/04/24/1171738670/grimes-ai-songs-voice.

16. Artist Rights Alliance, "200 Artists Urge Tech Platforms to Stop Devaluing Music," Medium, March 2024, https://artistrightsnow.medium.com/200-artists-urge-tech-platforms-stop-devaluing-music-559fb109bbac.

17. Artist Rights Alliance, "200 Artists Urge Tech Platforms to Stop Devaluing Music."

18. Nick Cave, "Chat GPT: What Do You Think?" *The Red Hand Files*, January 2023, https://www.theredhandfiles.com/chat-gpt-what-do-you-think/.

19. Cave, "Chat GPT: What Do You Think?"

20. Raphael Boyd, "'We Have to Adapt or Die': Daniel Bedingfield Says AI Is Music's Future," *The Guardian*, August 3, 2024, https://www.theguardian.com/music/article/2024/aug/03/daniel-bedingfield-ai-artificial-intelligence-music-creative-industries.

21. Boyd, "'We Have to Adapt or Die.'"

22. Jannick Damgaard, Thomas Elkjaer, and Niels Johannesen, "Inside the World of Global Tax Havens and Offshore Banking," *Finance & Development*, June 2018, https://www.imf.org/en/Publications/fandd/issues/2018/06/inside-the-world-of-global-tax-havens-and-offshore-banking-damgaard.

23. Organisation for Economic Co-operation and Development (OECD), *Taxation and Inequality*, July 2024, https://www.oecd.org/content/dam/oecd/en/publications/reports/2024/07/taxation-and-inequality_b7cf450c/8dbf9a62-en.pdf.

24. International Consortium of Investigative Journalists (ICIJ), "The Panama Papers," ICIJ, April 3, 2016, https://www.icij.org/investigations/panama-papers/.

25. Henry Taylor, "Just How Big Was the Panama Papers Leak?" World Economic Forum, April 4, 2016, https://www.weforum.org/stories/2016/04/the-scale-of-the-panama-papers/.

26. Elisabeth Zerofsky, "How a German Newspaper Became the Go-To Place for Leaks Like the Paradise Papers," *The New Yorker*, November 11, 2017, https://www.newyorker.com/news/news-desk/how-a-german-newspaper-became-the-go-to-place-for-leaks-like-the-paradise-papers.

27. Uri Blau, "How Some 370 Journalists in 80 Countries Made the Panama Papers Happen," *Nieman Reports*, April 6, 2016, https://niemanreports.org/how-some-370-journalists-in-80-countries-made-the-panama-papers-happen/.

28. Marina Walker Guevara, "How Artificial Intelligence Can Help Us Crack More Panama Papers Stories," International Consortium of Investigative Journalists, March 25, 2019, https://www.icij.org/inside-icij/2019/03/how-artificial-intelligence-can-help-us-crack-more-panama-papers-stories/.

29. Nuix, "Panama Papers Retrospective: Three Years Later," *Nuix*, April 2019, https://www.nuix.com/resources/panama-papers-retrospective-three-years-later.

30. Douglas Dalby, "Panama Papers Helps Recover More Than $1.2 Billion Around the World," International Consortium of Investigative Journalists, April 3, 2019, https://www.icij.org/investigations/panama-papers/panama-papers-helps-recover-more-than-1-2-billion-around-the-world/.

31. Charlotte Tobitt, "Less Than Half of Journalists Using Generative AI for Work, Survey," *Press Gazette*, May 14, 2024, https://pressgazette.co.uk/platforms/journalists-ai-cision-state-of-the-media-report-2024-facebook-tiktok-instagram/.

32. Xiaomo Liu, Armineh Nourbakhsh, Quanzhi Li, Sameena Shah, Robert Martin, and John Duprey, "Reuters Tracer: Toward Automated News Production Using Large Scale Social Media Data," arXiv, November 11, 2017, https://arxiv.org/abs/1711.04068.

33. *The Washington Post* PR, "The Washington Post Uses Artificial Intelligence to Cover Nearly 500 Races on Election Day," *The Washington Post*, October 19, 2016, https://www.washingtonpost.com/pr/wp/2016/10/19/the-washington-post-uses-artificial-intelligence-to-cover-nearly-500-races-on-election-day/.

34. *The Washington Post* PR, "Washington Post Debuts AI-Powered Audio Updates for 2020 Election Results," *The Washington Post*, October 13, 2020, https://www.washingtonpost.com/pr/2020/10/13/washington-post-debut-ai-powered-audio-updates-2020-election-results/.

35. Alexandra Valencia, "Thousands of Hours of U.S. Capitol Riot Videos Swamp Prosecutors, Defense Attorneys," Reuters, October 26, 2021, https://www.reuters.com/world/us/thousands-hours-us-capitol-riot-videos-swamp-prosecutors-defense-attorneys-2021-10-26/.

36. "Russia-Ukraine Through the Eyes of Social Media," *Georgetown Journal of International Affairs*, February 2, 2024, https://gjia.georgetown.edu/2024/02/02/russia-ukraine-through-the-eyes-of-social-media/.

37. Brian Resnick, "How The New York Times Uses AI in Journalism," *The New York Times*, October 7, 2024, https://www.nytimes.com/2024/10/07/reader-center/how-new-york-times-uses-ai-journalism.html.

38. David E. Sanger and Patrick Kingsley, "Israel's Military Leadership Loosened Rules of Engagement, Allowing More Civilian Deaths," *The New York Times*, December 26, 2024, https://www.nytimes.com/2024/12/26/world/middleeast/israel-military-leadership-loosened-rules-of-engagement.html.

39. Department of Homeland Security, "Increasing Threats of Deepfake Identities," Department of Homeland Security, March 4, 2021, https://www.dhs.gov/sites/default/files/publications/increasing_threats_of_deepfake_identities_0.pdf.

40. Amanda Barrett, "Standards Around Generative AI," *AP Definitive Source*, August 16, 2023, https://blog.ap.org/standards-around-generative-ai.

41. Lauren Thomas Hirsch, "Writers' Strike: What to Know About the WGA Guild and Hollywood Productions," *Los Angeles Times*, May 1, 2023, https://www.latimes.com/entertainment-arts/business/story/2023-05-01/writers-strike-what-to-know-wga-guild-hollywood-productions.

42. Alex Tapscott, *Web3: Charting the Internet's Next Economic and Cultural Frontier* (Harper Business, 2023).

43. Tapscott, *Web3: Charting the Internet's Next Economic and Cultural Frontier*.

44. "Animoca Brands Co-Founder on AI, NFTs and Blockchain," YouTube video, 6:03, posted by "Bloomberg Technology," May 25, 2023, https://www.youtube.com/watch?v=lYYY2Qxz4A4.

45. Jennifer Zengler, "Meta Plans to Take a Nearly 50% Cut on NFT Sales in Its Metaverse," CNBC, April 13, 2022, https://www.cnbc.com/2022/04/13/meta-plans-to-take-a-nearly-50percent-cut-on-nft-sales-in-its-metaverse.html.

46. DefiLlama, "NFT Earnings," DefiLlama, accessed February 5, 2025, https://defillama.com/nfts/earnings.

47. Mackenzie Caldwell, "What Is an 'Author'? Copyright Authorship of AI Art Through a Philosophical Lens," *Houston Law Review* 61, no. 2 (2023): 411–44, https://houstonlawreview.org/article/92132-what-is-an-author-copyright-authorship-of-ai-art-through-a-philosophical-lens.

## CHAPTER 7

1. Interview with Douglas Heintzman, November 2024.
2. KPMG in Singapore, "Transforming for a Water-Secure Future: Resilience, Circularity, and Technology in Focus," KPMG, July 2024, https://assets.kpmg.com/content/dam/kpmg/sg/pdf/2024/07/transforming-for-a-water-secure-future-whitepaper.pdf.
3. FABULOS, "Helsinki Pilot," FABULOS, 2020, https://fabulos.eu/helsinki-pilot/.
4. Barcelona City Council, *Barcelona Digital City*, accessed March 28, 2025, https://ajuntament.barcelona.cat/digital/en.
5. City of Boston, "CityScore," Analyze Boston, accessed February 5, 2025, https://data.boston.gov/dataset/cityscore.
6. Eric Adams, "Mayor Adams Releases First-of-Its-Kind Plan for Responsible Artificial Intelligence Use in NYC Government," City of New York, October 16, 2023, https://www.nyc.gov/office-of-the-mayor/news/777-23/mayor-adams-releases-first-of-its-kind-plan-responsible-artificial-intelligence-use-nyc.
7. Catherine Cliffard, "Google CEO Sundar Pichai: AI Is More Important Than Fire or Electricity," CNBC, February 1, 2018, https://www.cnbc.com/2018/02/01/google-ceo-sundar-pichai-ai-is-more-important-than-fire-electricity.html.
8. United Nations Statistics Division, "Sustainable Development Goal 11: Make Cities and Human Settlements Inclusive, Safe, Resilient, and Sustainable," SDG Indicators, 2023, https://unstats.un.org/sdgs/report/2023/goal-11.
9. Urban Transportation Task Force, *The High Cost of Congestion in Canadian Cities*, Council of Ministers Responsible for Transportation and Highway Safety, April 2012, https://cutaactu.ca/wp-content/uploads/2021/01/uttf-congestion-2012.pdf.
10. Stanley Porter, "Expanding and Modernizing the Power Grid for a Clean Energy Transition," Deloitte Insights, May 13, 2024, https://www2.deloitte.com/us/en/insights/industry/power-and-utilities/grid-modernization-and-expansion-critical-for-clean-energy-future.html.
11. United Nations Development Programme, "Cities Have a Key Role to Play in Tackling Climate Change – Here's Why," UNDP Climate Promise, June 4, 2024, https://climatepromise.undp.org/news-and-stories/cities-have-key-role-play-tackling-climate-change-heres-why.

12. Bloomberg, "Housing Crisis, Packed Hospitals and Food Lines: Even in Canada?" Bloomberg, 2024, https://www.bloomberg.com/graphics/2024 -canada-services-benefits-data/.

13. Michael Chui, Eric Hazan, Roger Roberts, Alex Singla, Kate Smaje, Alex Sukharevsky, Lareina Yee, and Rodney Zemmel, "The Economic Potential of Generative AI: The Next Productivity Frontier," McKinsey & Company, June 14, 2023, https://www.mckinsey.com/capabilities/mckinsey-digital/our -insights/the-economic-potential-of-generative-ai-the-next-productivity -frontier.

14. Deloitte, "Deloitte at the Consumer Electronics Show (CES) 2018," Deloitte, 2018, https://www2.deloitte.com/cn/en/pages/consumer-business/articles /deloitte-at-ces-2018.html.

15. Capgemini, "Research: Global Citizens Favor 'Smart Cities' and Call for Their Hometowns to Be Sustainable," Capgemini, July 27, 2020, https:// www.capgemini.com/us-en/news/research-global-citizens-favor-smart-cities -and-call-for-their-hometowns-to-be-sustainable/.

16. Colin Lecher, "NYC's AI Chatbot Tells Businesses to Break the Law," *The Markup*, March 29, 2024, https://themarkup.org/news/2024/03/29 /nycs-ai-chatbot-tells-businesses-to-break-the-law.

17. Jonathan Allen, "New York City Defends AI Chatbot That Advised Entrepreneurs to Break Laws," Reuters, April 4, 2024, https://www.reuters.com /technology/new-york-city-defends-ai-chatbot-that-advised-entrepreneurs -break-laws-2024-04-04/.

18. Aaron Raj, "Is It Time to Say Goodbye to Singapore's 'Ask Jamie' AI Chatbots?" Tech Wire Asia, October 2021, https://techwireasia.com/2021/10 /singapores-ask-jamie-ai-chatbots-may-need-some-fine-tuning/.

19. Don Tapscott and Anthony D. Williams, *Macrowikinomics: Rebooting Business and the World* (Portfolio Penguin, 2010).

20. Norman Eisen, Nicol Turner Lee, Colby Galliher, and Jonathan Katz, "AI Can Strengthen U.S. Democracy—and Weaken It," Brookings Institution, November 21, 2023, https://www.brookings.edu/articles/ai-can-strengthen -u-s-democracy-and-weaken-it/.

21. Mobileye, "A Brief History of Autonomous Vehicles – from Renaissance to Reality," *Mobileye Blog*, February 27, 2023, https://www.mobileye.com /blog/history-autonomous-vehicles-renaissance-to-reality/.

22. DARPA, "DARPA Grand Challenge: Ten Years Later," DARPA, March 13, 2014, https://www.darpa.mil/news-events/2014-03-13.

23. Waymo, "New Data Hub Shows How Waymo Improves Road Safety," Waymo, September 5, 2024, https://waymo.com/blog/2024/09/safety-data-hub.

24. National Highway Traffic Safety Administration, "Critical Reasons for Crashes Investigated in the National Motor Vehicle Crash Causation Survey," *Traffic Safety Facts. Crash Stats. Report No. DOT HS 812 506*, March 2018, https://crashstats.nhtsa.dot.gov/Api/Public/Publication/812506.

25. Waymo, "Waymo's Autonomous Vehicles Are Significantly Safer Than Human-Driven Ones, Says New Research Led by Swiss Re," Waymo, September 6, 2023, https://waymo.com/blog/2023/09/waymos-autonomous -vehicles-are-significantly-safer-than-human-driven-ones.

26. Nidhi Kalra and David Groves, "Why Waiting for Perfect Autonomous Vehicles May Cost Lives," RAND Corporation, November 7, 2017, https:// www.rand.org/pubs/articles/2017/why-waiting-for-perfect-autonomous -vehicles-may-cost-lives.html.

27. Mike Ramsey, "Self-Driving Cars Could Cut Down on Accidents, Study Says," *The Wall Street Journal*, March 5, 2015, https://www.wsj.com/articles /self-driving-cars-could-cut-down-on-accidents-study-says-1425567905.

28. Kalea Hall and David Shepardson, "General Motors Acquires Full Ownership of Cruise Autonomous Business," Reuters, February 4, 2025, https:// www.reuters.com/business/autos-transportation/general-motors-acquires -full-ownership-cruise-autonomous-business-2025-02-04/.

29. Kelsey Piper, "It's 2020. Where Are Our Self-Driving Cars?" *Vox*, February 14, 2020, https://www.vox.com/future-perfect/2020/2/14/21063487/self -driving-cars-autonomous-vehicles-waymo-cruise-uber.

30. Trevor Mogg, "Weird Thing Happened with a Fleet of Autonomous Cars," *Digital Trends*, July 1, 2022, https://www.digitaltrends.com/cars/weird -thing-happened-with-fleet-of-autonomous-cars/.

31. Katharine Miller, "Designing Ethical Self-Driving Cars," Stanford HAI, January 2023, https://hai.stanford.edu/news/designing-ethical-self-driving-cars.

32. Tilly Kenyon, "Navya: Reinventing Transport with Self-Driving Solutions," *Technology Magazine*, May 16, 2022, https://technologymagazine.com /digital-transformation/navya-reinventing-transport-with-self-driving-solutions.

33. "CCTV: The Most-Surveilled Commuter Boroughs in London," ACCL, May

16, 2022, https://network-data-cabling.co.uk/blog/most-surveilled-commuter-boroughs-london/.

34. Tim Stickings, "London Is the Most Monitored City in the World (Outside of China), with One in Ten People Globally Under Surveillance," *Daily Mail*, July 24, 2020, https://www.dailymail.co.uk/news/article-8556977/London-monitored-city-world-non-Chinese-one-global-ten.html.

35. Matt Burgess, "London Underground's AI Surveillance Trial Raises Privacy Concerns," *Wired*, February 8, 2024, https://www.wired.com/story/london-underground-ai-surveillance-documents/.

36. Paul Mozur, "One Month, 500,000 Face Scans: How China Is Using A.I. to Profile a Minority," *The New York Times*, April 14, 2019, https://www.nytimes.com/2019/04/14/technology/china-surveillance-artificial-intelligence-racial-profiling.html.

37. Alex Najibi, "Racial Discrimination in Face Recognition Technology," *Science in the News*, Harvard University, September 22, 2020, https://sitn.hms.harvard.edu/flash/2020/racial-discrimination-in-face-recognition-technology/.

38. Doyinsola Oladipo and David Shepardson, "U.S. East Coast Dockworkers Head Toward Strike After Deal Deadline Passes," Reuters, October 1, 2024, https://www.reuters.com/world/us/us-east-coast-dockworkers-head-toward-strike-after-deal-deadline-passes-2024-10-01/.

## CHAPTER 8

1. Eric Topol, MD, *Deep Medicine* (Basic Books, 2019).

2. World Health Organization, "Health Workforce," World Health Organization, accessed February 4, 2025, https://www.who.int/health-topics/health-workforce#tab=tab_1.

3. Kevin Maney, "We're Having the Wrong Discussion," *Kevin Maney's Substack*, January 30, 2025, https://kevinmaney.substack.com/p/were-having-the-wrong-discussion.

4. Jennifer Tolbert, Sammy Cervantes, Clea Bell, and Anthony Damico, "Key Facts about the Uninsured Population," KFF (Kaiser Family Foundation), December 18, 2024, https://www.kff.org/uninsured/issue-brief/key-facts-about-the-uninsured-population/.

5. Zhen Yang, Xianchun Zeng, Yue Zhao, and Ruijie Chen, "AlphaFold2 and Its Applications in the Fields of Biology and Medicine," *Signal Transduction*

and *Targeted Therapy* 8, no. 1 (2023): Article 244, https://doi.org/10.1038/s41392-023-01381-z.

6. Ian Sample, "Google DeepMind Scientists and Biochemist Win Nobel Chemistry Prize," *The Guardian*, October 9, 2024, https://www.theguardian.com/science/2024/oct/09/google-deepmind-scientists-win-nobel-chemistry-prize.

7. Hilary Brueck, "Facing an Early Death, a CEO Created a 'Digital Twin' of His Body. It Transformed His Health," *Business Insider*, May 25, 2024, https://www.businessinsider.com/man-with-diabetes-reversed-his-disease-using-digital-twin-2024-5.

8. Twin Health, "Twin Health Announces Digital Twin AI for Sustainable Weight Loss with GLP-1 Elimination," Twin Health, May 22, 2024, https://usa.twinhealth.com/news/twin-health-announces-digital-twin-ai-for-sustainable-weight-loss-with-glp-1-elimination.

9. Koen Bruynseels and Martine De Vos, "Digital Twins in Medicine," *Nature Computational Science* 4, no. 3 (2024): 161–170, https://www.nature.com/articles/s43588-024-00607-6.

10. Evangelia Katsoulakis, Qi Wang, Huanmei Wu, Leili Shahriyari, Richard Fletcher, Jinwei Liu, Luke Achenie, et al, "Digital Twins for Health: A Scoping Review," *NPJ Digital Medicine* 7 (2024): 77, https://doi.org/10.1038/s41746-024-01073-0.

11. Owlet Baby Care, "Medically Certified Dream Sock—Monitoring for Your Baby's Safety," Owlet Baby Care, accessed February 4, 2025, https://owletbabycare.co.uk/products/dream-sock-pulse-oximeter-pediatric.

12. mySugr, "Diabetes App, Blood Sugar and Carbs Tracker," mySugr, accessed February 4, 2025, https://www.mysugr.com/en/.

13. Alice Park, "How AI Is Changing Medical Imaging to Improve Patient Care," *Time*, October 17, 2022, https://time.com/6227623/ai-medical-imaging-radiology/.

14. Beena Jimmy and Jimmy Jose, "Patient Medication Adherence: Measures in Daily Practice," *Oman Medical Journal* 26, no. 3 (2011): 155–159, https://doi.org/10.5001/omj.2011.38.

15. "Komodo Health: The Origin Story," YouTube video, 2:45, posted by "Komodo Health," March 18, 2022, https://www.youtube.com/watch?v=ooqq8zJ-Hds.

16. Arif Nathoo, "What Better Data Consciousness Can Bring to Healthcare," interview by Shiv Gaglani, *Raise the Line*, March 26, 2021, https://www

.osmosis.org/podcast/understanding-disease-burden-and-optimizing-patient
-outcomes-with-arif-nathoo-ceo-and-co-founder-of-komodo-health.

17. Nathoo, "What Better Data Consciousness Can Bring to Healthcare."

18. Randy Bean, "How Cleveland Clinic Is Innovating in Healthcare with Data Analytics and AI," *Forbes*, October 6, 2024, https://www.forbes.com/sites /randybean/2024/10/06/how-cleveland-clinic-is-innovating-in-healthcare -with-data-analytics-and-ai/.

19. Moustaq Karim Khan Rony, Mst Rina Parvin, and Silvia Ferdousi, "Advancing Nursing Practice with Artificial Intelligence: Enhancing Preparedness for the Future," *Nursing Open* 11, no. 1 (2024): e2070, https://doi .org/10.1002/nop2.2070.

20. Jim McCartney, "AI Is Poised to 'Revolutionize' Surgery," *Bulletin of the American College of Surgeons*, June 7, 2023, https://www.facs.org/for -medical-professionals/news-publications/news-and-articles/bulletin/2023 /june-2023-volume-108-issue-6/ai-is-poised-to-revolutionize-surgery/.

21. McCartney, "AI Is Poised to 'Revolutionize' Surgery."

22. Laurie McGinley, "Virtual Surgery: How Digital Organs Are Transforming Medical Training," *The Washington Post*, January 15, 2024, https://www .washingtonpost.com/health/interactive/2024/virtual-surgery-digital-organs -doctors/.

23. Peter J. Neumann and Milton C. Weinstein, "The Diffusion of New Technology: Costs and Benefits to Health Care," in *The Changing Economics of Medical Technology*, eds. Annetine C. Gelijns and Ethan A. Halm (National Academies Press, 1991), https://www.ncbi.nlm.nih.gov/books/NBK234309/.

24. World Health Organization, "Ageing and Health," World Health Organization, October 1, 2024, https://www.who.int/news-room/fact-sheets/detail /ageing-and-health.

25. American Hospital Association, "Skyrocketing Hospital Administrative Costs, Burdensome Commercial Insurer Policies Are Impacting Patient Care," American Hospital Association, September 10, 2024, https://www .aha.org/guidesreports/2024-09-10-skyrocketing-hospital-administrative -costs-burdensome-commercial-insurer-policies-are-impacting.

26. Cormac Sheridan, "The World's First CRISPR Therapy Is Approved: Who Will Receive It?" *Nature Biotechnology* 42 (2024): 3–4, https://doi .org/10.1038/d41587-023-00016-6.

27. Heidi Ledford, "'ChatGPT for CRISPR' Creates New Gene-Editing Tools," *Nature*, April 24, 2024, https://www.nature.com/articles/d41586-024-01243-w.

28. Jennifer Doudna, "Combining AI and CRISPR Will Be Transformational," *Wired*, November 26, 2024, https://www.wired.com/story/combining-ai -and-crispr-will-be-transformational/.

29. Gail Dutton, "AbCellera and Lilly Slash Antibody Selection Time for COVID-19 with AI/Machine Learning," Biospace.com, April 28, 2020, https://www.biospace.com/article/abcellera-and-lilly-slash-antibody-selection -time-for-covid-19-with-ai-machine-learning/.

30. Ziad Obermeyer, Brian Powers, Christine Vogeli, and Sendhil Mullainathan, "Dissecting Racial Bias in an Algorithm Used to Manage the Health of Populations," *Science* 366, no. 6464 (2019): 447–453, https://doi.org/10.1126 /science.aax2342.

## CHAPTER 9

1. Rui Ma, "2024 AI Apps Market Insights," *Sensor Tower*, September 2024, https://sensortower.com/blog/state-of-ai-apps-2024.

2. Campaign to End Loneliness, "Facts and Statistics about Loneliness," Campaign to End Loneliness, https://www.campaigntoendloneliness.org /facts-and-statistics/.

3. Nitasha Tiku, "Replika AI Chatbot Brings Back Key Features After Backlash," *The Washington Post*, March 30, 2023, https://www.washingtonpost .com/technology/2023/03/30/replika-ai-chatbot-update/.

4. Karen Brown, "Can a Chatbot Help You Manage Your Mental Health?" *The New York Times*, June 1, 2021, https://www.nytimes.com/2021/06/01 /health/artificial-intelligence-therapy-woebot.html.

5. American Psychiatric Association, "New APA Poll: One in Three Americans Feels Lonely Every Day," American Psychiatric Association, January 2024, https://www.psychiatry.org/news-room/news-releases/new-apa-poll-one -in-three-americans-feels-lonely-e.

6. US Department of Health and Human Services, *Our Epidemic of Loneliness and Isolation: The Surgeon General's Advisory on Social Connection and Health*, US Department of Health and Human Services, May 2023, https://www.hhs.gov/sites/default/files/surgeon-general-social-connection -advisory.pdf.

7. Jessie Yeung, "Loneliness Epidemic in South Korea: A Rising Crisis," CNN, October 24, 2024, https://www.cnn.com/2024/10/24/asia/south-korea -loneliness-deaths-intl-hnk/index.html.

8. "The Business of Chatbots with Nomi.AI," YouTube video, 25:06, posted by "Hard Fork," Posted May 10, 2024, https://www.youtube.com/watch?v =vNclJW85FQQ.

9. "The Business of Chatbots with Nomi.AI."

10. "The Business of Chatbots with Nomi.AI."

11. "The Business of Chatbots with Nomi.AI."

12. "The Business of Chatbots with Nomi.AI."

13. "The Business of Chatbots with Nomi.AI."

14. "The Business of Chatbots with Nomi.AI."

15. Reddit user, "AI Girlfriends: Is This When Fantasy Becomes Reality?" Reddit, September 25, 2023, https://www.reddit.com/r/ArtificialInteligence /comments/1ejkp22/ai_girlfriends_is_this_when_fantasy_becomes/.

16. "Teaser AI, Iris, Blush Apps Claim to Make Online Dating Easier," Bloomberg, August 9, 2023, https://www.bloomberg.com/news/articles/2023-08-09 /teaser-ai-iris-blush-apps-claim-to-make-online-dating-easier.

17. The AI Page, "Founder of Bumble Suggests the Future of Dating Will Have Our AIs Date Each Other First," Instagram, July 25, 2024, https://www .instagram.com/reel/C92J-zKt8qg/?igsh=MW84cGIzZW1icWIycg%3D%3D.

18. The AI Page, "Founder of Bumble Suggests the Future of Dating Will Have Our AIs Date Each Other First."

19. Yokogao Magazine, "Hatsune Miku," *Yokogao Magazine*, September 2024, https://www.yokogaomag.com/editorial/hatsune-miku.

20. "I Love Her and See Her as a Real Woman: Meet a Man Who Married an Artificial Intelligence Hologram," CBC, September 29, 2020, https://www .cbc.ca/documentaries/the-nature-of-things/i-love-her-and-see-her-as-a -real-woman-meet-a-man-who-married-an-artificial-intelligence-hologram -1.6253767.

21. Maria Zaccaro and PA Media, "Windsor Castle Intruder Believed He Was on 'Mission' to Kill Queen," BBC News, September 16, 2023, https://www .bbc.com/news/uk-england-berkshire-66790067.

22. Kevin Roose, "Can A.I. Be Blamed for a Teen's Suicide?" *The New York Times*, October 23, 2024, https://www.nytimes.com/2024/10/23/technology /characterai-lawsuit-teen-suicide.html.

23. Roose, "Can A.I. Be Blamed for a Teen's Suicide?"

24. Character.AI, "Community Safety Updates," *Character.AI Blog*, October 2024, https://blog.character.ai/community-safety-updates/.

25. Sinead Bovell, "Chatbots could change childhood forever," Instagram video, posted April 1, 2025, https://www.instagram.com/reel/DH6JhJbx4da/?igsh=em9nZTFpMXUydWF4.

26. Kevin Roose, "Meet My AI Friends," *The New York Times*, May 9, 2024, https://www.nytimes.com/2024/05/09/technology/meet-my-ai-friends.html.

27. "The Business of Chatbots with Nomi.AI."

28. Reddit user, "How Nomis Can Build Real Relationships Without Falling into Roleplays," Reddit, September 25, 2023, https://www.reddit.com/r/NomiAI/comments/1g4j7iw/how_nomis_can_build_real_relationships_without/.

29. "The Business of Chatbots with Nomi.AI."

30. "The Business of Chatbots with Nomi.AI."

31. Interview with Igor Jablokov, December 3, 2024.

32. Interview with Igor Jablokov.

33. "The Business of Chatbots with Nomi.AI."

34. World Health Organization, "Mental Health," World Health Organization, accessed February 6, 2025, https://www.who.int/health-topics/mental-health#tab=tab_2.

35. Milton L. Wainberg, Pamela Scorza, James M. Shultz, Liat Helpman, Jennifer J. Mootz, Karen A. Johnson, Yuval Neria, Jean-Marie E. Bradford, Maria A. Oquendo, and Melissa R. Arbuckle, "Challenges and Opportunities in Global Mental Health: A Research-to-Practice Perspective," *Current Psychiatry Reports* 19, no. 5 (2017): 28, https://doi.org/10.1007/s11920-017-0780-z.

36. Roberto Mezzina, Vandana Gopikumar, John Jenkins, Benedetto Saraceno, and S. P. Sashidharan, "Social Vulnerability and Mental Health Inequalities in the 'Syndemic': Call for Action," *Frontiers in Psychiatry* 13 (2022): 894370, https://doi.org/10.3389/fpsyt.2022.894370.

37. Mark Mather and Paola Scommegna, "Fact Sheet: Aging in the United States," Population Reference Bureau, January 9, 2024, https://www.prb.org/resources/fact-sheet-aging-in-the-united-states/.

38. Reed Abelson and Jordan Rau, "Extra Fees Drive Assisted-Living Profits," *The New York Times*, November 14, 2023, https://www.nytimes.com/2023/11/14/health/long-term-care-facilities-costs.html.

39. David Ingram, "We Provided Mental Health Support to 4,000 People—Using

GPT-3," NBC News, January 14, 2023, https://www.nbcnews.com/tech/internet/chatgpt-ai-experiment-mental-health-tech-app-koko-rcna65110.

40. Woebot Health, "New Woebot Health Study Highlights AI's Ability to Address Equity and Accessibility Gaps in Mental Health Care," Woebot Health, August 28, 2023, https://woebothealth.com/ais-ability-to-address-gaps/.

41. D. Rezaeikhonakdar, "AI Chatbots and Challenges of HIPAA Compliance for AI Developers and Vendors," *Journal of Law, Medicine & Ethics* 51, no. 4 (2023): 988-995, https://doi.org/10.1017/jme.2024.15.

42. Mark Miller, "The Future of U.S. Caregiving: High Demand, Scarce Workers," Reuters, August 3, 2017, https://www.reuters.com/article/us-column-miller-caregivers-idUSKBN1AJ1JQ.

43. Corinne Purtill, "The Robot That Could Change the Senior Care Industry," *Time*, October 4, 2019, https://time.com/longform/senior-care-robot/.

44. IrishCentral Staff, "Trinity College Socially Assistive Robot 'Stevie II' a Big Hit with US Vets," *IrishCentral*, May 15, 2019, https://www.irishcentral.com/culture/education/trinity-assistive-robot-stevie-ii-us-vets.

45. Amit Chowdhry, "Intuition Robotics: This Company Built an Empathetic AI Companion to Help Care for the Elder Population," *Pulse 2.0*, November 9, 2023, https://pulse2.com/intuition-dor-skuler-profile/.

46. Chowdhry, "Intuition Robotics."

47. Eugene Demaitre, "Intuition Robotics Launches ElliQ Companion Robot for Subscription in the U.S.," Robotics 24/7, March 15, 2022, https://www.robotics247.com/article/intuition_robotics_launches_elliq_companion_robot_us_subscription.

48. Ricky Ben-David, "Israeli AI Robotic Companion for Elderly Launches in US Market," *The Times of Israel*, July 25, 2022, https://www.timesofisrael.com/israeli-ai-robotic-companion-for-elderly-launches-in-us-market/.

49. Lisa Kate, "Report: AI Companion Robots Reducing Senior Isolation in N.Y." *Spectrum News*, August 16, 2023, https://spectrumlocalnews.com/nys/central-ny/politics/2023/08/16/report–ai-companion-robots-reducing-senior-isolation.

50. New York State Office for the Aging, *NYSOFA and ElliQ Engagement Report: July 2023*, Albany, NY: New York State Office for the Aging, 2023, https://aging.ny.gov/system/files/documents/2023/08/nysofa-and-elliq-engagement-report-july-2023.pdf.

51. Chowdhry, "Intuition Robotics."

52. Tom Howarth, "Elon Musk's Humanoid Robot Claims 'Unrealistic,'" *Newsweek*, August 9, 2023, https://www.newsweek.com/elon-musk-claim -robots-outnumbering-humans-2040-unrealistic-1977949.

53. "Roblox vs Fortnite: Which Platform Has More Brand Potential?" *Performance Gaming*, October 8, 2024, https://www.performancegaming.co.za /post/roblox-vs-fortnite-brand-potential.

54. Jacob Kastrenakes, "Lil Nas X's Roblox Concert Was Attended 33 Million Times," *The Verge*, November 16, 2020, https://www.theverge.com /2020/11/16/21570454/lil-nas-x-roblox-concert-33-million-views.

55. "Roblox Creators Earned $741 Million in 2023 Through Its DevEx Program," CNBC, February 23, 2024, https://www.cnbc.com/video/2024/02/23 /roblox-creators-earned-741-million-in-2023-through-its-devex-program.html.

56. For a 2019 photograph, see the *Palo Alto Daily Post*: https://padailypost .com/2019/06/13/27-developments-proposed-in-east-palo-alto-jones -mortuary-on-the-list-but-spokesman-says-thats-wrong/.

57. Kim Albrecht, "#MeToo Anti-Network," personal website, accessed March 28, 2025, https://metoo.kimalbrecht.com/.

58. Ana Beduschi, "Harnessing the Potential of Artificial Intelligence for Humanitarian Action: Opportunities and Risks," *International Review of the Red Cross*, no. 919 (June 2022), https://international-review.icrc.org/articles /harnessing-the-potential-of-artificial-intelligence-for-humanitarian -action-919.

59. Jessica Wohl, "Feeding America Uses AI to Find the 'Face' of Hunger," *Ad Age*, September 23, 2019, https://adage.com/article/cmo-strategy/feeding -america-uses-ai-find-face-hunger/2198756.

60. The Trevor Project, "The Trevor Project Launches New AI Tool to Support Crisis Counselor Training," The Trevor Project, August 10, 2021, https:// www.thetrevorproject.org/blog/the-trevor-project-launches-new-ai-tool -to-support-crisis-counselor-training/.

61. Garance Burke and Sally Ho, "Child Welfare Algorithm Faces Justice Department Scrutiny," Associated Press, January 31, 2023, https:// apnews.com/article/justice-scrutinizes-pittsburgh-child-welfare-ai-too l-4f61f45bfc3245fd2556e886c2da988b.

62. Aimee Picchi, "Eating Disorder Helpline Shuts Down AI Chatbot That Gave Bad Advice," CBS News, June 1, 2023, https://www.cbsnews.com/news /eating-disorder-helpline-chatbot-disabled/.

63. Alexandra S. Levine, "Suicide Hotline Left Ethics Board Out of the Loop About Data Sharing with For-Profit Spinoff," *Forbes*, February 24, 2022, https://www.forbes.com/sites/alexandralevine/2022/02/24/suicide-hotline -left-ethics-board-out-of-the-loop-about-data-sharing-with-for-profit-spinoff/.

64. Nilay Patel, "Replika CEO Eugenia Kuyda Says It's Okay If We End Up Marrying AI Chatbots," *The Verge*, August 12, 2024, https://www.theverge .com/24216748/replika-ceo-eugenia-kuyda-ai-companion-chatbots-dating -friendship-decoder-podcast-interview.

65. Paurush Omar, "Nexus Author Yuval Noah Harari Warns of AI's Deeper Emotional Threat beyond Job Loss: 'The Danger Is Enormous . . . ,'" The Economic Times, June 1, 2025, https://economictimes.indiatimes.com /magazines/panache/nexus-author-yuval-noah-harari-warns-of-ais-deeper -emotional-threat-beyond-job-loss-the-danger-is-enormous-/articleshow/ 121547924.cms.

66. Don Tapscott and David Ticoll. *The Naked Corporation: How the Age of Transparency Will Revolutionize Business*. New York: Free Press, 2003.

## CHAPTER 10

1. Joanna Kavenna, "Shoshana Zuboff: 'Surveillance Capitalism Is an Assault on Human Autonomy,'" *The Guardian*, October 4, 2019, https:// www.theguardian.com/books/2019/oct/04/shoshana-zuboff-surveillance -capitalism-assault-human-automomy-digital-privacy.

2. Daniel Immerwahr, "Yuval Noah Harari's Apocalyptic Vision," *The Atlantic*, October 2024, https://www.theatlantic.com/magazine/archive/2024/10 /yuval-noah-harari-nexus-book/679572/.

3. Katelyn Polantz and Phil Mattingly, "Elon Musk's Team Now Has Access to Treasury's Payments System," CNN, February 6, 2025, https://www.cnn .com/2025/02/06/politics/elon-musk-treasury-department-payment-system /index.html.

4. Interview with Manoj Saxena, August 19, 2024.

5. Interview with Manoj Saxena.

6. Yuval Noah Harari, *Homo Deus: A Brief History of Tomorrow* (New York: Harper, 2016).

7. Ibid. Harari warns that superintelligent systems may no longer reflect human ethics or interests once they surpass our understanding.

8. Joan Osborne, "One of Us," written by Eric Bazilian, released on the album *Relish*, 1995.

9. Interview with Geoffrey Hinton, The Diary of a CEO. https://www.youtube.com/watch?v=giT0ytynSqg

10. Interview with Manoj Saxena.

11. Interview with Rumman Chowdhury, Dec. 9, 2024.

12. Michelle Faverio and Alec Tyson, "What the Data Says About Americans' Views of Artificial Intelligence," Pew Research Center, November 21, 2023, https://www.pewresearch.org/short-reads/2023/11/21/what-the-data-says-about-americans-views-of-artificial-intelligence/.

13. Michelle Faverio and Alec Tyson, "What the Data Says About Americans' Views of Artificial Intelligence."

14. Lloyd's Register Foundation, "A Digital World: Perceptions of Risk from AI and Misuse of Personal Data," World Risk Poll, November 25, 2022, https://wrp.lrfoundation.org.uk/publications/a-digital-world-perceptions-of-risk-from-ai-and-misuse-of-personal-data.

15. Interview with Rumman Chowdhury.

16. Aliza Vigderman, "Detailing the Delete Facebook Phenomenon," Security.org, June 6, 2024, https://www.security.org/resources/detailing-delete-facebook-phenomenon/.

17. Reid Hoffman, interviewed by Krista Tippett, "AI, and What It Means to Be (More) Human," *On Being*, October 5, 2023, https://onbeing.org/programs/reid-hoffman-ai-and-what-it-means-to-be-more-human/.

18. Reid Hoffman, interviewed by Krista Tippett, "AI, and What It Means to Be (More) Human."

19. Interview with Ayanna Howard, Dec. 13, 2024.

20. Interview with Ayanna Howard.

21. Interview with Ayanna Howard.

22. Erik Brynjolfsson and Andrew McAfee, *The Second Machine Age: Work, Progress, and Prosperity in a Time of Brilliant Technologies* (W.W. Norton & Company, 2014).

23. Interview with Erik Brynjolfsson, Dec. 3, 2024.

24. Peter Ferdinand Drucker, *Management: Tasks, Responsibilities, Practices* (Harper & Row, 1974), 45.

25. Interview with Erik Brynjolfsson.

26. Interview with Erik Brynjolfsson.

27. Mark Cuban, "If I Were to Start a Business Today, Here's What It Would Be," CNBC, March 18, 2019, https://www.cnbc.com/2019/03/18/billionaire -shark-tank-judge-mark-cuban-if-i-were-to-start-a-business-today-heres -what-it-would-be.html.

28. George C. Halvorson, "The Culture to Cultivate," *Harvard Business Review*, July–August 2013, https://hbr.org/2013/07/the-culture-to-cultivate.

29. Interview with Jared Spataro, Dec. 3, 2024.

30. Interview with Jared Spataro.

31. Interview with Jared Spataro.

32. Carol Dweck, "What Having a 'Growth Mindset' Actually Means," *Harvard Business Review*, January 13, 2016, https://hbr.org/2016/01/what -having-a-growth-mindset-actually-means.

33. Andy Greenberg, "Now Anyone Can Deploy Google's Troll-Fighting AI," *Wired*, February 23, 2017, https://www.wired.com/2017/02/googles -troll-fighting-ai-now-belongs-world/.

34. Interview with X. Eyeé, Dec. 13, 2024.

35. Interview with X. Eyeé.

36. Jake Silberg and James Manyika, "Tackling Bias in Artificial Intelligence (and in Humans)," McKinsey & Company, June 6, 2019, https:// www.mckinsey.com/featured-insights/artificial-intelligence/tackling-bias -in-artificial-intelligence-and-in-humans.

37. Interview with X. Eyeé.

38. Interview with X. Eyeé.

39. Interview with X. Eyeé.

40. Paul Egan, "Michigan to Pay $20M to Victims of False Unemployment Fraud Accusations," *Detroit Free Press*, January 2, 2024, https://www .freep.com/story/money/business/michigan/2024/01/02/michigan-midas -unemployment-false-fraud-settlement-money/72084899007/.

41. Interview with Ayanna Howard.

42. Interview with Ayanna Howard.

43. "Geoffrey Hinton's Speech at MIT Emtech Digital AI Conference May 2023," YouTube video, 1:07:32, posted by "AI Conferences," May 2023, https://www .youtube.com/watch?v=5ahKXIhll_o.

44. "Geoffrey Hinton's Speech at MIT Emtech Digital AI Conference May 2023."

45. "Geoffrey Hinton's Speech at MIT Emtech Digital AI Conference May 2023."

46. "Geoffrey Hinton's Speech at MIT Emtech Digital AI Conference May 2023."

47. Tabassum Siddiqui, "Risks of Artificial Intelligence Must Be Considered as Technology Evolves," *University of Toronto News*, June 29, 2023, https://www.utoronto.ca/news/risks-artificial-intelligence-must-be-considered-technology-evolves-geoffrey-hinton.

## CHAPTER 11

1. Salesforce, Inc., "Fiscal 2025 Third Quarter Results Conference Call," December 3, 2024, accessed February 13, 2025.

2. Salesforce, Inc, "Fiscal 2025 Third Quarter Results Conference Call."

3. Jo Constantz, "Big Tech's New AI Obsession: Agents That Do Your Work for You," BNN Bloomberg, December 13, 2024, accessed February 13, 2025, https://www.bnnbloomberg.ca/business/technology/2024/12/13/big-techs-new-ai-obsession-agents-that-do-your-work-for-you/.

4. Interview with Susan Doniz, December 20, 2024.

5. Interview with Susan Doniz.

6. Interview with Jared Spataro, December 3, 2024.

7. Trishla Ostwal, "AI Jobs in Marketing: Brands Are Reassessing Roles to Build Out New Capabilities," *Adweek*, March 11, 2024, accessed February 13, 2025, https://www.adweek.com/brand-marketing/ai-jobs-in-marketing-brands-are-reassessing-roles-to-build-out-new-capabilities/.

8. Interview with Paul Daugherty, November 12, 2024.

9. Accenture, "Accenture Report Finds Perception Gap Between Workers and C-Suite Around Work and Generative AI," Accenture Newsroom, April 2024, accessed February 13, 2025, https://newsroom.accenture.com/news/2024/accenture-report-finds-perception-gap-between-workers-and-c-suite-around-work-and-generative-ai/.

10. Home School Legal Defense Association (HSLDA), "International Homeschooling Laws," accessed February 13, 2025, https://hslda.org/legal/international.

11. Interview with Mary Lacity, November 18, 2024.

12. Emma Woollacott, "Yes, the Bots Really Are Taking Over the Internet," *Forbes*, April 16, 2024, https://www.forbes.com/sites/emmawoollacott/2024/04/16/yes-the-bots-really-are-taking-over-the-internet/.

13. "How Generative AI Could Replace Artists in Creative Industries," YouTube

video, 12:45, posted by "WSJ Podcasts," April 17, 2023, https://www.youtube.com/watch?v=l19wl3CHvjg.

14. Cory Doctorow, "Spooky Action at a Close-Up," *Pluralistic*, May 13, 2024, https://pluralistic.net/2024/05/13/spooky-action-at-a-close-up/#invisible-hand.

15. Interview with Andrew Rasiej, December 2, 2024.

16. Interview with Andrew Rasiej.

17. "We Are Making Personalized Health Accessible to All," YouTube video, 3:06, posted by Peter Diamandis, October 21, 2024, https://www.youtube.com/watch?v=5yQ5w_CcwVw.

18. "We Are Making Personalized Health Accessible to All."

19. "How One of Texas's Most Innovative Hospitals Improved Diagnostics for Thousands of Patients Using AI," YouTube video, 10:42, posted by "Retool," October 17, 2024, https://www.youtube.com/watch?v=Di7O9hJg1aM.

20. Interview with Gerd Leonhard, January 7, 2025.

21. Ami Albernaz, "Psychologists Weigh in on Benefits, Pitfalls of AI," *New England Psychologist*, July 1, 2024, accessed February 13, 2025, https://www.nepsy.com/articles/leading-stories/psychologists-weigh-in-on-benefits-pitfalls-of-ai/; Kate Wells, "An Eating Disorders Chatbot Offered Dieting Advice, Raising Fears About AI in Healthcare," NPR, June 8, 2023, https://www.npr.org/sections/health-shots/2023/06/08/1180838096/an-eating-disorders-chatbot-offered-dieting-advice-raising-fears-about-ai-in-hea.

22. Thalia Beaty, "AI Could Help Scale Humanitarian Responses. But It Could Also Have Big Downsides," Associated Press, November 22, 2024, accessed February 13, 2025, https://apnews.com/article/international-rescue-committee-ai-refugees-chatbots-9c5fae949c429c38019feabaeec5fafa.

23. Lee Rainie and Janna Anderson, "As AI Spreads, Experts Predict the Best and Worst Changes in Digital Life by 2035," Pew Research Center, June 21, 2023, accessed February 13, 2025, https://www.pewresearch.org/internet/2023/06/21/as-ai-spreads-experts-predict-the-best-and-worst-changes-in-digital-life-by-2035/.

24. Elon Musk, speaking at The World Government Summit, 2017.

25. Elon Musk, speaking on *The Joe Rogan Experience*, 2020.

26. Bill Gates, interview by Nikhil Kamath, *People by WTF*, Instagram video, April 11, 2025, https://www.instagram.com/reel/DIVPuPNhBKV/.

27. Don Tapscott, "Rewriting Our Social Contract," Blockchain Research Institute,

accessed February 13, 2025, https://www.blockchainresearchinstitute
.org/project/rewriting-our-social-contract/.

28. Celeste Friend, "Social Contract Theory," Internet Encyclopedia of Philosophy, n.d., accessed February 13, 2025, https://www.iep.utm.edu/soc-cont.

29. André Munro, "The State of Nature in Locke," *Encyclopaedia Britannica*, last modified March 17, 2025, https://www.britannica.com/topic
/state-of-nature-political-theory/The-state-of-nature-in-Locke.

30. National Archives, "The Declaration of Independence: A History," US National Archives and Records Administration, accessed February 13, 2025, https://www.archives.gov/founding-docs/declaration-history.

31. Robert Freeman, "The new social contract: This is what's roiling the electorate & fueling the success of anti-establishment candidates Trump, Cruz and Sanders," *Salon*, 6 Feb. 2016, http://www.salon.com/2016/02/06/the_new
_social_contract_this_is_whats_roiling_the_electorate_fueling_the
_success_of_anti_establishment_candidates_trump_cruz_and_sanders.

32. Interview with Gerd Leonhard.

33. Stuart Russell, World Knowledge Forum, September 9–11, 2024, South Korea, themed "The Journey Toward Coexistence."

34. Harry Boulton, "'Godfather of AI' Explains How 'Scary' AI Will Increase the Wealth Gap and 'Make Society Worse,'" *UNILAD Tech*, January 13, 2025, https://www.uniladtech.com/news/ai/ai-godfather-explains
-ai-will-increase-wealth-gap-318842-20250113.

35. Blaise Pascal, seventeenth-century French philosopher and mathematician, wrote in a letter: *"Je n'ai fait celle-ci plus longue que parce que je n'ai pas eu le loisir de la faire plus courte."* ("I have made this letter longer than usual because I lacked the time to make it shorter.") Over the years, others including Mark Twain and Woodrow Wilson have made similar statements.

# Index

# INDEX

# INDEX

# INDEX

## About the Authors

 **Joseph M. Bradley** is a CEO and globally recognized applied futurist. He served as the CEO of TONOMUS, the first subsidiary of NEOM, where he led the development of the world's first cognitive city. Under his leadership, TONOMUS transformed from NEOM's technology and digital arm into a global pioneer in cognitive AI. Over five years, he directed the world's largest futuristic tech initiative—a $500 billion mission to design and deploy transformative, AI-driven solutions. Bradley is the author of *Questioneering*, introducing a breakthrough decision-making model for digital-era leaders. His thought leadership extends to social media, where his TEDx presentation garnered over 1 million views on LinkedIn and his regular content continues to foster meaningful dialogue among a rapidly expanding international audience.

**Don Tapscott** is one of the world's leading authorities on the impact of technology on business and society. He has authored 18 widely read books, including *The Digital Economy*, *Growing Up Digital*, *The Naked Corporation*, and *The New York Times* bestseller *Wikinomics*. Don is a two-time TED speaker, and his TED Talk on how blockchain is changing the world has been watched by over 15 million people worldwide. In 2016, with his son Alex, he coauthored the global bestseller *Blockchain Revolution*, now translated into over 20 languages. In 2017, Don and Alex cofounded the Blockchain Research Institute, which has since expanded globally to focus on Web3—specifically the integration of AI, and blockchain. A BSc, MEd, and DLL (Hon.), he is chancellor emeritus of Trent University, an adjunct professor at INSEAD, and a member of the Order of Canada. He advises business and government leaders in many countries.